"*Systems Medicine* is a masterpiece. Written in a conversational style, it invites us to think about health and disease in a profoundly simple new way. From the secrets of aging to the enigma of autoimmune diseases, Uri Alon uses physiological circuits and the principles that underlie them to illuminate how our bodies work, why specific diseases occur, and what strategies might be used to treat them. Whether you're a curious mind, a biology enthusiast, or just someone excited to understand the magnificent orchestra of life, this book has something extraordinary to offer. To my mind, it's nothing short of revolutionary."

—**Steven Strogatz**, *Cornell University, USA, author of Nonlinear Dynamics and Chaos*

"Uri Alon has once again opened our eyes to a new frontier of quantitative thinking, this time applied to medicine. *Systems Medicine* is a masterpiece of clear interdisciplinary exposition, suitable for interested students and researchers with minimal mathematical or physiological background. Its inspiring narrative, accompanied by engaging exercises, explains how the recent application of dynamical systems methods to diabetes, stress, ageing, and autoimmune disorders resolves long-standing puzzles in medicine. This book is destined to be as influential as his earlier *Systems Biology*, along with potentially important impacts on human health."

—**Nigel Goldenfeld**, *University of California San Diego, USA*

"This unique book will benefit an unusually broad range of students, scientists, and medical professionals. If you have a background in math, physics, or engineering, you will learn key concepts in physiology and medicine presented in the familiar language of dynamical systems. If your background is in biomedical sciences, you will learn how to build and employ powerful mathematical models to unveil hidden patterns behind disease vulnerabilities. Either way, Uri Alon will masterfully guide you to new insights into the underlying logic of physiological systems and their susceptibility to diseases."

—**Ruslan Medzhitov**, *Yale University, USA*

"This book will be as influential and transformative as Alon's *Systems Biology*. Its quantitative approach provides new insight into the mechanisms that underline common diseases. It should be of interest to students and researchers in medicine, biology, and engineering. It will introduce quantitative analysis to medical students, interesting medical problems to engineers, and catalyze a new synthesis of "systems medicine". Alon is a pioneer and leading expert in systems biology. He is now pioneering a new field with this book."

—**Liqun Luo**, *Stanford University, USA*

"This book is a paradigm-shifting journey elegantly guiding readers to perceive health and disease through a new lens. By weaving together multiple disciplines, from mathematics to biology and medicine, Alon paints a vivid picture of how our understanding of biological networks and their dynamics can revolutionize medical approaches. Using simple concepts coupled with real-world examples, the book offers a comprehensive exploration of physiological circuits and exemplifies the practical applications of Systems Medicine. With insightful clarity, Alon distills intricate scientific concepts into digestible insights, making this book accessible to both seasoned researchers and curious minds new to the field."

—**Galit Lahav**, *Harvard University, USA*

Systems Medicine

Why do we get certain diseases, whereas other diseases do not exist?

In this book, Alon, one of the founders of systems biology, builds a foundation for systems medicine.

Starting from basic laws, the book derives why physiological circuits are built the way they are. The circuits have fragilities that explain specific diseases and offer new strategies to treat them.

By the end, the reader will be able to use simple and powerful mathematical models to describe physiological circuits. The book explores, in three parts, hormone circuits, immune circuits, and aging and age-related disease. It culminates in a periodic table of diseases.

Alon writes in a style accessible to a broad range of readers - undergraduates, graduates, or researchers from computational or biological backgrounds. The level of math is friendly and the math can even be bypassed altogether. For instructors and readers who want to go deeper, the book includes dozens of exercises that have been rigorously tested in the classroom.

ABOUT THE AUTHOR

Uri Alon is the Abisch-Frenkel Professor of Systems Biology at the Weizmann Institute of Science, Israel. https://www.weizmann.ac.il/mcb/UriAlon/homepage

Chapman & Hall/CRC
Computational Biology Series

About the Series

This series aims to capture new developments in computational biology, as well as high-quality work summarizing or contributing to more established topics. Publishing a broad range of reference works, textbooks, and handbooks, the series is designed to appeal to students, researchers, and professionals in all areas of computational biology, including genomics, proteomics, and cancer computational biology, as well as interdisciplinary researchers involved in associated fields, such as bioinformatics and systems biology.

For more information about this series please visit: https://www.routledge.com/Chapman—HallCRC-Computational-Biology-Series/book-series/CRCCBS

Systems Medicine
Physiological Circuits and the Dynamics of Disease

Uri Alon
Weizmann Institute of Science, Israel

CRC Press
Taylor & Francis Group
Boca Raton London New York

CRC Press is an imprint of the
Taylor & Francis Group, an **informa** business

A CHAPMAN & HALL BOOK

Designed cover image: Nigel Orme

First edition published 2024
by CRC Press
2385 NW Executive Center Drive, Suite 320, Boca Raton, FL 33431

and by CRC Press
4 Park Square, Milton Park, Abingdon, Oxon, OX14 4RN

CRC Press is an imprint of Taylor & Francis Group, LLC

© 2024 Uri Alon

Library of Congress Cataloging-in-Publication Data
Names: Alon, Uri, 1969- author.
Title: Systems medicine : physiological circuits and the dynamics of disease / Uri Alon.
Description: Boca Raton : CRC Press, 2024. | Includes bibliographical references and index. | Contents: The insulin-glucose circuit – Dynamical compensation, mutant resistance, and type-2 diabetes – The stress hormone axis as a two-gland oscillator – Autoimmune diseases as a fragility of mutant surveillance – Inflammation and fibrosis as a bistable system – Basic facts of aging – Aging and saturated repair – Age-related diseases – Periodic table of diseases – Epilogue: Simplicity in systems medicine.
Identifiers: LCCN 2023014141 (print) | LCCN 2023014142 (ebook) | ISBN 9781032412283 (hardback) | ISBN 9781032411859 (paperback) | ISBN 9781003356929 (ebook)
Subjects: LCSH: Biological systems–Mathematical models. | Human physiology–Mathematical models. | Systems biology.
Classification: LCC QH324.2 .A467 2024 (print) | LCC QH324.2 (ebook) | DDC 570.285–dc23/eng/20230630
LC record available at https://lccn.loc.gov/2023014141
LC ebook record available at https://lccn.loc.gov/2023014142

ISBN: 978-1-032-41228-3 (hbk)
ISBN: 978-1-032-41185-9 (pbk)
ISBN: 978-1-003-35692-9 (ebk)

DOI: 10.1201/9781003356929

Typeset in Minion
by codeMantra

Contents

CHAPTER 3.5 ■ The Thyroid and Its Discontents 76

PART II **Immune Circuits**

CHAPTER 4 ■ Autoimmune Diseases as a Fragility of Mutant Surveillance 89

Introduction

H<small>I, I'M</small> U<small>RI</small> A<small>LON</small>, <small>A PROFESSOR</small> from the Weizmann Institute in Israel. In my PhD in physics I looked for patterns in turbulent mixing; what I loved about physics were the moments of seeing, in complex systems, suddenly an angle where things looked simple.

Toward the end of my PhD, I looked for subjects where the physics way of thinking might help to find new laws of nature. I didn't know much about biology – the only thing I knew about proteins was what I read on the back of a cereal box. Then a friend gave me a textbook on cell biology. Here was matter that was alive! I fell in love with biology and resolved to see if there are laws to be found.

I wasn't alone. Luckily, Stanislas Leibler took me on as a postdoc, where I met other physicists who shared the vision of biological principles, Naama Barkai and Michael Elowitz. I was encouraged by biologists like Arnold Levine, Yosef Yarden, and Benny Geiger. Before long, I had my own group back in Israel.

In the first decade, we focused on understanding the cell, with its networks of interacting proteins. At the time, around 2000, there was massive information on which protein interacts with whom, but it was hard to make sense of this information: the networks of interactions looked hopelessly complex. We discovered that these networks are simpler than they appear. They are made of a handful of recurring basic circuits which we named network motifs. These motifs show up again and again in different systems and in all organisms, and each has its own computational function. The basic circuits inside the cell are described in my book *An Introduction to Systems Biology*.

Then, 10 years ago, I saw a poster in the elevator in my building that changed everything. The poster announced a talk by Yuval Dor, saying that glucose makes the cells that control it, called beta cells, both grow and die. This is a paradox – why does glucose do two opposite things to the same cells? It reminded me of a paradox I had studied in bacteria, where enzymes do a reaction and also its reverse. I felt I could do something in an exciting field – human hormone circuits.

This was an opening to a new phase in my research career. I fell in love again with human medicine and physiology, and how physics-style thinking can help make sense of our bodies in health and illness.

DOI: 10.1201/9781003356929-1

I'm excited to start this book with you, on systems medicine.

My wish is that some of you will feel the same way – that you can do something in this field – and join us.

Our topic is physiological circuits that describe how cells and organs communicate with each other. Rather than circuits inside a cell, we will discuss circuits of communication between cells. This level of description is relevant to some of the most common and deadly diseases that plague humanity.

It's good to think about the goal of the book. **The goal is to start from basic principles or laws and derive why physiology is built the way it is, why specific diseases happen, and which new strategies might treat them.**

By the end, you will be able to use simple and powerful mathematical models to describe physiological circuits. The models are powerful because they turn details into useful understanding and new ways to think about medicine. We will understand the fundamental causes of some of the most mysterious diseases: diabetes, autoimmune diseases, and age-related diseases such as lung fibrosis and cancer.

Our trajectory begins with basic principles. From these, we will derive the circuits and their fragility to disease. We will explore, in three parts, hormone circuits, immune circuits, and aging and age-related disease. Our story culminates in a periodic table of diseases.

ABOUT MATH AND BIOLOGICAL TERMS

I write this book with a heterogeneous readership in mind. Your background might be biology, engineering, physics, math, computer science, medicine, or other subjects, as an undergraduate, graduate, or researcher.

For some of you the following equation is familiar, whereas others need brushing up:

$$\frac{dx}{dt} = -\alpha x$$

We will use this equation to describe the removal of cells, whose number is x. The rate of removal is α; you can think about this as the probability per unit time that a cell dies. Since cells are only removed in this equation, and not added, their number declines. In fact, $x(t)$ declines exponentially with time, t, starting from their initial number $x(0)$, as given by the solution

$$x(t) = x(0)e^{-\alpha t}$$

You can check this solution by taking the derivative $\frac{dx}{dt}$, and since a derivative of an exponential $e^{-\alpha t}$ is $-\alpha e^{-\alpha t}$, you get our equation back, $\frac{dx}{dt} = -\alpha x$.

That's the level of the math in the book, ordinary differential equations that describe changes over time. You can skip the equations and still enjoy the book or go into them and learn ways of thinking about modeling.

Similarly, for some readers, biological terms like beta cells and hormones are familiar; for others, they are new. I'll assume no prior knowledge and use minimal jargon – biologists and physicians will see much fewer gene names than they are used to. I'll explain terms when needed. A hormone is a molecule secreted by a set of cells into the blood, where it reaches and affects cells in distant parts of the body; beta cells are the cells that secrete the hormone insulin.

I'd also like to say what this book is not. It is not a book on medical bioinformatics, the gathering and statistical interpretation of biomedical data. This field, sometimes also called systems medicine, is described in books listed at the end of this chapter. We will not discuss applications of machine learning or artificial intelligence to medicine. We will, however, use large medical datasets in this book to test our mechanistic models. This book is also not exhaustive, and I provide further reading below. The purpose of the book is to provide a way of thinking, and the examples are chosen to clearly demonstrate principles.

OUR FIRST FEEDBACK LOOP

This book is also written to be fun. So let's jump right in! Here is our first feedback loop (Figure 0.1). A person can be in a relaxed state of mind. The relaxed state is good for learning and memory. In the relaxed state, our body behaves in specific ways. For example, we take slow deep breaths.

The wonderful thing is that we can decide to take a deep breath, and this increases the chances that we enter the relaxed state. Because the relaxed state is good for learning, we will practice taking nice deep sighs of relief in this book from time to time. Let's practice now: you don't have to, but if you do, I promise you will enjoy it. Let's all together take a nice deep sigh of relief.

The next chapter gives a taste of our approach and teaches basic concepts within a fascinating bit of physiology. So let's take a nice deep sigh of relief – here we go!

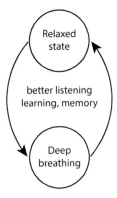

FIGURE 0.1   The relaxed state is better for listening, learning, and remembering.

BACKGROUND READING

This book combines several disciplines – Systems Biology, Evolutionary Medicine, Mathematical Physiology and Dynamical Systems. These disciplines are exemplified by the following books which are recommended reading if you want to go deeper.

Stearns, Stephen C., and Ruslan Medzhitov. 2015. *Evolutionary Medicine*. 1st edition. Sunderland, Massachussetts: Sinauer Associates is an imprint of Oxford University Press.

Strogatz, Steven H. 2015. *Nonlinear Dynamics and Chaos*: With Applications to Physics, Biology, Chemistry, and Engineering, Second Edition. Boulder, CO.

Keener, James, and James Sneyd, eds. *Mathematical physiology*: II: Systems physiology. New York, NY: Springer New York, 2009.

Alon, Uri. *An introduction to systems biology: design principles of biological circuits*. CRC press, 2019.

A different approach to systems medicine focuses on statistical analysis of large datasets. It is described in the following books.

Wolkenhauer, Olaf. *Systems Medicine: Integrative, Qualitative and Computational Approaches*. Academic Press, 2020.

Schmitz, Ulf, and Olaf Wolkenhauer, eds. *Systems Medicine*. Springer, 2016.

Loscalzo, Joseph, ed. *Network Medicine*. Harvard University Press, 2017.

Yan, Qing. *Translational Bioinformatics and Systems Biology Methods for Personalized Medicine*. Academic Press, 2017.

Bai, J.P.F. and J. Hur (Editors). *Systems Medicine*, 2022.

I

Hormone Circuits

CHAPTER 1

The Insulin-Glucose Circuit

IN THIS CHAPTER AND the next, we focus on the glucose-insulin circuit. It is the hydrogen atom of hormone circuits – it provides principles that apply to many other systems. The circuit is also important medically. Its failure is the basis for diabetes, a disease afflicting about 10% of the world's population (of which 90% is type-2 and 10% type-1 diabetes, stay tuned).

GLUCOSE CONCENTRATION AND DYNAMICS IS TIGHTLY CONTROLLED

The main variable in this system is the concentration of the sugar glucose in the blood. Glucose is an energy and carbon source for the cells in our body. It is the major fuel for the brain and for immune cells. **Glucose concentration in the blood is maintained within a tight range around 5 mM.** That reads '5 millimolar', or, in other common units, 90 mg per deciliter.

Glucose steady-state concentrations vary by only about 20% between healthy individuals. Such rigorous control is called **homeostasis** in biology – the ability of the body to keep important variables within a tight range. If glucose drops below 3 mM, the brain does not have enough energy and we can faint. Prolonged low glucose, called **hypoglycemia**, can be fatal. The body switches to alternative energy sources such as ketone bodies which can cause blood acidity, which is potentially lethal.

Similarly, if glucose is too high, above 7 mM, it starts sticking to blood vessels and damaging them. Over years, this leads to the deadly symptoms of type-2 diabetes. The damaged blood vessels cause heart attacks, kidney disease, and, in the retina, blindness. Damaged blood vessels can also lead to amputation of legs and other grim outcomes.

Let's take a nice deep sigh of relief.

The control is so rigorous that clinical criteria for diabetes are based on glucose blood tests. Blood glucose below 5.6 mM (100 mg/dL) after fasting for 8 hours or more is normal. Glucose between 5.6 and 6.9 mM is **prediabetes**, and above 6.9 mM (125 mg/dL) on two separate tests means diabetes.

In addition to the stringent control of the steady-state level of glucose, **the entire glucose dynamics after a meal is tightly regulated.** These dynamics are measured in a clinical test

DOI: 10.1201/9781003356929-3

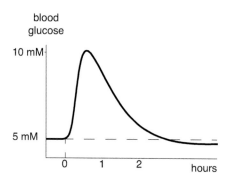

FIGURE 1.1 Blood glucose rises and falls over 2 hours after a glucose tolerance test.

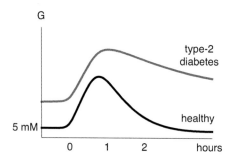

FIGURE 1.2 High glucose levels and slow recovery are clinical signs of diabetes in the glucose tolerance test.

for diabetes called the glucose tolerance test (GTT). In GTT, you drink 75 g of glucose and measure blood glucose in the following 2 hours. Glucose rises to about twice its basal level of 5 mM, and then falls back to baseline in about 2 hours (Figure 1.1). Different healthy people have similar glucose dynamics in the GTT. Aberrant dynamics are a sign of diabetes: glucose above 11 mM at 2 hours is a clinical criterion for diabetes (Figure 1.2).

We can now ask how such rigorous control is achieved despite vast differences between people. Individuals can vary in weight by a large factor, undergo the changes of pregnancy, vary in activity and diet, and so on. Yet most people most of the time have glucose set-points and dynamics that are nearly the same.

GLUCOSE CONCENTRATION IS CONTROLLED BY INSULIN

How is the control of blood glucose concentration achieved? The answer is a feedback circuit involving the hormone **insulin**, a small protein that circulates in the blood. Insulin acts to remove glucose from the blood, by allowing it to enter cells in the muscle, liver, and fat where glucose is used or stored.

Glucose is unable to enter these cells without special glucose transporters on the cell surface. The transporters are stored away in storage vesicles inside the cell (Figure 1.3A). When insulin is in the blood, it binds special sensors on the cell surface called insulin

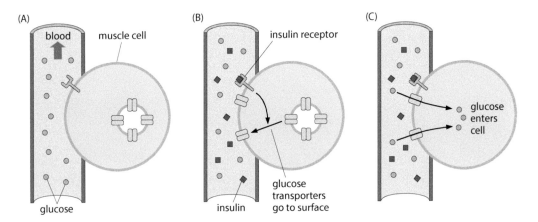

FIGURE 1.3   Insulin causes glucose to be removed from the blood into muscle and fat cells. Binding of insulin to its receptors causes transporters to translocate to the cell membrane and transport glucose into the cell. (A–C) described in text.

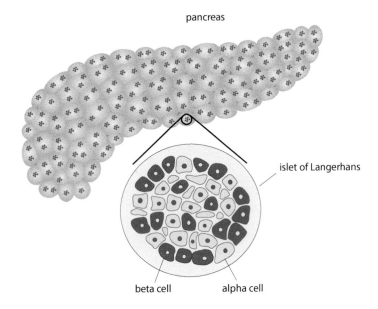

FIGURE 1.4   Beta cells which secrete insulin are found in Islets of Langerhans in the pancreas along with other hormone secreting cells such as alpha cells that secrete glucagon.

receptors (Figure 1.3B). These receptors bind insulin like a lock and key. When bound, the receptors initiate signaling pathways inside the cell that move the glucose transporters to the cell surface (Figure 1.3B), where they let glucose into the cell. As a result, insulin shunts glucose out of the blood and into the cells (Figure 1.3C).

Insulin is secreted by special cells in the pancreas called **beta cells**. The pancreas is a thin gland about the size of a dollar bill located in our upper abdomen (Figure 1.4). In this gland there are a billion beta cells, arranged in a million groups of about 1000 cells called islets of Langerhans. The islets also house other types of cells, like alpha cells that secrete

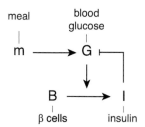

FIGURE 1.5 The glucose-insulin feedback loop, in which glucose causes beta cells to secrete insulin, which in turn removes glucose from the blood.

glucagon, a hormone that acts to increase glucose in the blood – countering the action of insulin. Glucagon induces the production of glucose in the liver during fasting and sleep.

Beta cells are smart, and only secrete insulin when it is needed. The beta cells sense glucose, and the more glucose around, the more insulin they secrete. Insulin induces cells in the muscle and fat to take up glucose, and so blood glucose levels drop. This is a negative feedback loop: more glucose means more insulin and thus less glucose (Figure 1.5). Note the blunt-headed arrow, a symbol of a negative regulation in the world of biological circuits – in this case insulin reducing glucose levels.

The input to this circuit is glucose from meals that goes into the blood. Between meals glucose is produced by the liver. The liver stores glucose in times of plenty in a polymer called glycogen and breaks it down when we fast. When it runs out of glycogen, at about the middle of the night, the liver makes glucose out of amino acids taken from muscles, in a process called gluconeogenesis ("new production of glucose"). We denote both sources by *m* in Figure 1.5.

DIABETES IS A MALFUNCTION IN THIS SYSTEM

In type-1 diabetes (T1D), the immune system attacks beta cells and kills them off. As a result, there is no insulin and glucose cannot enter muscle cells. Until the 1920s, type-1 diabetes was a death sentence to the children who got it: about 1% of the population gets T1D, typically at the age of 10–14. With the discovery of insulin, diabetic children survive thanks to insulin injections, and now through continuous insulin pumps. Still, keeping glucose under external control is hard, and type-1 diabetes raises the risk for health complications. We will understand this autoimmune disease in Chapter 4.

In the more common disease called type-2 diabetes (T2D), glucose rises over the years and causes damage to the body. A major cause of type-2 diabetes is insulin resistance, which we will describe below.

We have now completed a verbal introduction to this system. It is a basic version of the verbal description generally taught to physicians and biologists. The verbal description is powerful in that it can intuitively explain the dynamics, such as the rise and fall after a glucose tolerance test, and the basic phenomena in diabetes.

In this book, we want to go beyond verbal descriptions by adding equations. Equations can help us focus on important parameters and to generalize principles from one system to other systems. Most importantly, **equations help us to ask new questions**, such as what is the fundamental origin of diseases such as T1D and T2D. In this chapter, we lay the foundation for the next chapter in which we will make progress on these questions.

MATHEMATICAL MODEL FOR THE GLUCOSE-INSULIN CIRCUIT

Mathematical models for this circuit, developed since the 1970s, have benefitted clinical practice. They help to define key parameters like insulin resistance. They also provide practical ways to estimate these parameters for each patient based on clinical measurements. One important model is the minimal model by Richard Bergman (Bergman et al. 1979), and we will use a version of this model as a basis for our exploration.

The model was developed using experiments on volunteers in which glucose or insulin were introduced intravenously, and their effects were monitored over time using glucose and insulin blood tests. The model uses differential equations to describe rates of change of glucose and insulin concentrations in the blood.

This model, and all the models in this book, ignores some of the details of the system and focuses only on the main features essential to understand the principles we are interested in. The simplicity of the models might generate mistrust because one might think "It is more complicated than that! You didn't take this or that into account." We will see that ignoring some details is not only okay but necessary to see underlying principles. By exploring the art of building good models, we will learn how to tell which features are essential based on their timescale and magnitude.

Let's begin with the equation for the rate of change of insulin concentration, $I(t)$. The equation has a form we will use throughout the book, where the rate of change equals production minus removal:

$$\text{Rate of change of insulin} = dI/dt = \text{production-removal}$$

For more details, see Solved Exercise 1.7.

Let's start with insulin production. Insulin is produced by beta cells, and the production rate rises with glucose. Thus, each beta cell makes $q\,f(G)$ units of insulin concentration per unit time, where q is the maximal production rate per unit biomass of beta cells divided by the blood volume, and $f(G)$ is an increasing function of glucose G, that ranges between 0 and 1. It describes how glucose regulates the secretion rate of insulin.[1] Experimental measurements show an increasing S-shaped curve that saturates at high glucose, meaning that it reaches a maximal level (Figure 1.6; Alcazar and Buchwald 2019). Such curves are well described in many biological systems by a Hill function, given by

(1.1)
$$f(G) = \frac{G^n}{K^n + G^n}$$

This function reaches half of its maximum value at a glucose concentration of $G = K$. This half-way concentration is about $K = 8\,\text{mM}$ in human islets. The steepness of the Hill

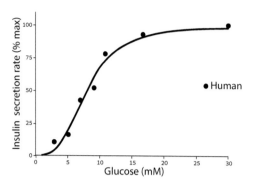

FIGURE 1.6 Insulin secretion by beta cells is a sigmoidal function of glucose. Adapted from Alcazar and Buchwald (2019).

function is controlled by the parameter n, the Hill coefficient. The higher n, the steeper the curve. For beta cells, $n=2$ is a good approximation (Figure 1.6). Hill functions have a theoretical basis which can be found in most biochemistry textbooks (see Solved Exercise 1.8) (Phillips et al. 2012).

All that we need to do now is to multiply the production rate by the total beta cell mass B to get an insulin production rate of $q\ B\ f(G)$.

We now turn to insulin removal. Removal of molecules is well described by giving them a constant rate of removal, γ, so that the number of molecules removed per unit time is γI. The difference between production and removal is our equation for insulin:

(1.2)
$$dI/dt = q\ B\ f(G) - \gamma I$$

The removal rate γ is determined by the insulin half-life, about 5 minutes. Insulin is removed primarily by degradation in the liver, and the remainder is filtered out by the kidneys.

To get some practice with the equation, let's see how the insulin removal rate γ relates to its half-life. Imagine blocking insulin production (as in patients with T1D who have no beta cells, or by using certain drugs). In this case, there is no production, only removal, and hence

$$dI/dt = -\gamma I$$

The solution of this equation is a concentration that decays exponentially with time from its initial level $I(0)$:

(1.3)
$$I(t) = I(0) \exp(-\gamma t)$$

The **half-life** of insulin, $t_{1/2}$, is the time it takes to go halfway down from its initial level. Thus $I(t_{1/2}) = I(0)/2$. Plugging this into Eq. 1.3, we find $\exp(-\gamma\, t_{1/2}) = 1/2$, and our solution for the half-life

$$t_{1/2} = \ln(2)/\gamma$$

This is a general result: **the half-life is inversely related to the removal rate** – faster removal leads to shorter half-life. Half-life is not affected by any of the production parameters like q and $f(G)$ or by initial conditions. Removal rates will similarly determine half-lives throughout this book.

Now let's write the second equation, for glucose. Blood glucose concentration, $G(t)$, is produced by meals and by liver production of glucose, whose sum is denoted $m(t)$. Glucose is removed by the action of insulin. The rate of change of glucose is, as before, supply minus removal

(1.4)
$$\frac{dG}{dt} = m - aG$$

Let's focus on removal term $-aG$. Imagine that you are an engineer that needs to design this circuit. The simplest design is not to have insulin at all, but rather to have a constant glucose removal rate, a, the probability per unit time to lose a glucose molecule from the blood. Since high levels of glucose are harmful, it makes sense to remove it quickly after meals, with a large removal rate a. However, rapid removal means that at night or during fasting, the liver would need to make more glucose per unit time to keep 5 mM steady state.

To see this, notice that the steady state of Eq. 1.4, that is, when glucose does not change with time, $dG/dt = 0$, is $G_{st} = m/a$. Steady-state glucose is the ratio of production and removal rates. The higher the removal rate a, the larger production m needs to be to maintain a 5 mM steady-state glucose concentration. Thus, a rapid glucose removal rate creates a wasteful cycle of high production and high removal.

To avoid this wasteful cycle, the body uses insulin to increase the removal rate of glucose when needed. Insulin is made only when glucose is high. This provides a high removal rate after a meal and a low removal rate during fasting. Notably, since insulin is made in tiny amounts compared to glucose, the cost of the insulin system itself is negligible.

Because glucose removal is enhanced by insulin, we let its removal rate rise with insulin, $a = s\,I$. The parameter s is called **insulin sensitivity**. It is an important parameter. Insulin sensitivity is the effect of a unit of insulin on glucose removal rate. This parameter is more familiar in its inverse form, **insulin resistance**, defined as $1/s$. Insulin resistance is the extent to which insulin fails to work. Insulin sensitivity can be measured by injecting insulin and observing the subsequent reduction in glucose.

Thus, our glucose equation is:

(1.5)
$$dG/dt = m - s\,I\,G$$

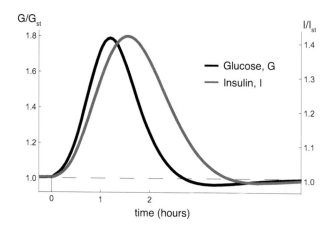

FIGURE 1.7 The minimal model captures the rise and fall of glucose and insulin over hours after a glucose input.

The insulin and glucose equations reach a steady state, which we can find by setting all rates of change to zero, namely $dG/dt=0$ and $dI/dt=0$. Solving this precisely yields a messy formula, but a good approximation can be made by assuming that $f(G)=(G/K)^2$, as described in Solved Exercise 1.1. The steady-state glucose concentration is

$$(1.6) \qquad G_{st} = (\gamma K^2 m_{st}/s\ q\ B)^{1/3}$$

The 5 mM steady state can be achieved provided the parameters are right.

Let's see how these equations do in the glucose tolerance test. We can solve the equations on the computer and provide a pulse of input glucose $m(t)$ to describe the glucose going into the body when we drink 75 g of glucose. In response, $G(t)$ rises, causing insulin $I(t)$ to rise in turn, thereby increasing the removal rate of G, until it returns to baseline, as shown in Figure 1.7 (see Exercise 1.6 for more details). This resembles the measured response of healthy people, including a slight glucose undershoot before returning to baseline. The minimal model thus seems to capture the essential behavior.

A GRAPHICAL TOOL TO UNDERSTAND THE CIRCUIT, THE PHASE PORTRAIT

Often, we don't know for sure what is the precise mathematical model to describe a biological circuit. I'd like to present a way in which we can still make progress. This is a graphical tool that we will employ throughout the book, called the **phase portrait** (Strogatz 2018). The phase portrait will show us that the feedback loop can reach a steady state no matter what the exact parameters or forms of the interaction terms are. Later, we will use it to see how parameters like insulin resistance affect the system.

The phase portrait has two axes, one for each of the variables, in our case insulin and glucose. The idea is to break the feedback loop into two arms. Study each arm separately, and then put them back together.

The first arm describes how glucose induces beta cells to make insulin (black line in Figure 1.8). One experiment to measure this arm of the feedback loop uses intravenous

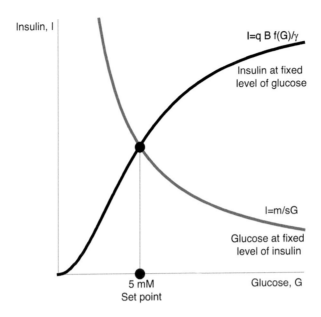

FIGURE 1.8 The feedback loop can be broken down into two arms called nullclines that meet at the fixed point.

injection of glucose to "clamp" blood glucose to a given value. Then, one allows insulin to reach steady state and measures its concentration. Plotting insulin versus glucose reveals a rising S-shaped function. The more glucose, the more insulin secreted by the beta cells.

In the model, this line is given by setting $dI/dt=0$ in Eq. 1.2, corresponding to zero change in insulin. This is called the insulin **nullcline**. Its formula is $I=qBf(G)/\gamma$, and so the nullcline has essentially the same shape as the induction function $f(G)$.

The second arm of the feedback loop (blue line in Figure 1.8) describes how insulin removes glucose. You clamp blood insulin at a given level and wait till glucose settles down to its steady state. At each level of insulin, you record the steady-state level of glucose, and plot this to get the decreasing blue line in Figure 1.7. The curve decreases because more insulin removes more glucose from the blood. Its equation can be derived by setting $dG/dt=0$ in Eq. 1.5, to get the glucose nullcline, $I=m/s\,G$.

Now we unleash the full feedback loop without artificially clamping one of the variables. Glucose and insulin now affect each other. The point to watch is where the two lines intersect. This is the **fixed point** of the system, where both glucose and insulin are at steady state simultaneously. These are the values you can measure with blood tests after fasting, say in the morning.

We can add to the phase portrait little arrows that indicate the dynamics (Figure 1.9). The arrows show the direction of change for insulin and glucose. They converge into the fixed point, showing that all initial conditions flow to it – it is a **stable fixed point**. The orange trajectory in Figure 1.9 shows how glucose and insulin evolve when we start with an initial excess of glucose. Glucose makes insulin rise, pushing glucose down toward its steady state of 5 mM, and both return to baseline.

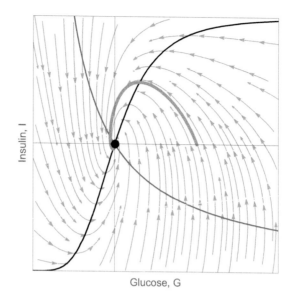

Glucose, G

FIGURE 1.9 The phase portrait shows arrows that indicate insulin and glucose dynamics. One trajectory is shown in orange which begins with high glucose and flows into the fixed point.

The phase portrait can help us arrive at conclusions even if we don't know the mathematical functions or parameters in the model and hence the precise shapes of the nullclines. No matter what the exact shape of the curves, we know that one curve must increase – glucose induces insulin production – and the other curve must decrease – insulin removes glucose. This guarantees a steady state where the two curves intersect.

So far so good. The minimal model provides a glucose set point, and the observed rise and fall of glucose and insulin in response to a meal. We can take a nice deep sigh of relief for making it this far.

INSULIN SENSITIVITY VARIES WIDELY BETWEEN PEOPLE

With our model in hand, let's now ask about the tightness of glucose regulation. A striking observation is that one of the model parameters varies widely between individual people, but empirical evidence shows that the glucose steady state and dynamics remain constant. Is our current model capable of capturing this kind of consistency?

The varying parameter is insulin sensitivity, s. Insulin sensitivity varies between people and over time because it is a physiological knob that allows the body to allocate glucose resources and determine which tissues get the glucose. For example, when we exercise or during caloric restriction, we need to use or store glucose, and signals are secreted that cause insulin sensitivity, s, to rise. The effect of insulin is magnified by higher s, and muscles and fat take up more glucose from the blood.

In contrast, during infection and inflammation, insulin sensitivity drops – insulin resistance rises – due to inflammatory signals in the circulation. Instead of storing it in muscle and fat, more glucose stays in the blood to help the immune system fight pathogens. Insulin resistance also rises during pregnancy, diverting glucose to the fetus. A further condition that leads to insulin resistance is chronic stress, through the action of the hormone cortisol

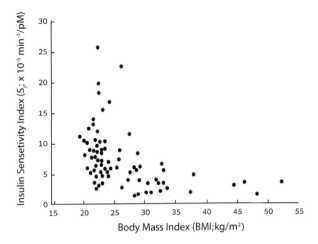

FIGURE 1.10   Insulin sensitivity drops with body mass index. Adapted from Kahn et al. (1993).

that we will discuss in Chapter 3. A common cause of insulin resistance is obesity – *s* drops dramatically, often by a factor of 10, with body mass index (weight divided by height squared) as shown in Figure 1.10. Obesity causes fat to accumulate in muscle and liver, triggering inflammation and insulin resistance (James, Stöckli, and Birnbaum 2021).

Thus inflammation, pregnancy, chronic stress, and obesity lead to insulin resistance (low *s*). Each unit of insulin works less effectively than in non-resistant people. Insulin resistance, as we will see in the next chapter, is an important factor in T2D.

Even though people vary in *s* by as much as a factor of ten, most people have normal glucose levels and dynamics. For example, most people with obesity, which all have low *s*, have normal 5 mM glucose and GTT dynamics.

THE MINIMAL MODEL FAILS TO EXPLAIN HOW PEOPLE WITH INSULIN RESISTANCE MAINTAIN NORMAL GLUCOSE LEVELS

Let's see what the model predicts for insulin resistance. Suppose that *s* drops by a factor of 10 – or equivalently insulin resistance rises by a factor of 10, making insulin 10 times less effective at removing glucose. In the phase portrait (Figure 1.11), the glucose nullcline shifts to a higher level, because that nullcline is inversely proportional to *s*, $I = m/sG$. As a result, the glucose set-point shifts to higher levels, far above 5 mM.

The phase portrait shows that this is a general effect, no matter what the exact shape of the curves. Insulin resistance shifts the glucose nullcline up because a given insulin level results in more glucose, and hence steady-state glucose rises.

This creates a problem for the model because all people with obesity have insulin resistance, but most have normal glucose levels.

Indeed, if we simulate the minimal model with a 10-fold lower *s*, we see that steady-state glucose concentration rises by a factor of about 2 (Figure 1.12 blue line is the model prediction). You can see this from the solution of Eq. 1.6, where G_{st} depends on $1/s^{\frac{1}{3}}$, and the cube root of 10 is about 2. In fact, glucose steady state depends on all model parameters.

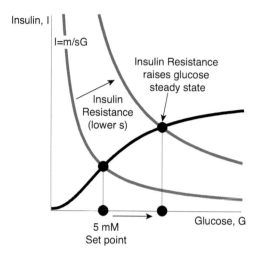

FIGURE 1.11   Insulin resistance raises one nullcline, shifting the fixed point to high glucose levels.

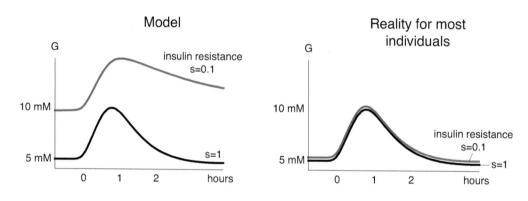

FIGURE 1.12   Insulin resistance in the minimal model causes high glucose dynamics, whereas most people with insulin resistance have normal glucose dynamics.

The response time after a glucose tolerance test also increases substantially in the model when s is low (Figure 1.12 blue line). See Solved Exercise 1.2 for an analytic solution for the response time, which also depends on all model parameters.

Thus, the minimal model cannot explain how most people with obesity have normal glucose (their observed dynamics is like that predicted for $s=1$, black line). In fact, no model based on the description of the system we studied so far can do so. We need to add another control loop to make glucose dynamics robust to variations in parameters such as s. We will do this in the next chapter.

SUMMARY AND OUTLOOK

The prose description of the insulin-glucose circuit found in textbooks seems to work qualitatively well. But when we write the equations, we see that we need additional mechanisms to explain how glucose stays under tight control despite variations in physiological parameters. We need to explain why most people with obesity, pregnancy, or chronic stress have very different insulin resistances but normal 5 mM glucose and normal dynamics in the glucose tolerance test.

In the next chapter, we will see how answering this question opens up general principles for feedback control in tissues. This new feedback has unavoidable fragilities that explain why beta cells fail in type-2 diabetes, as we will see in the next chapter, and why the body attacks its own beta cells in type-1 diabetes, as we will see later in Chapter 4. In the next chapter, we will also reveal the three basic "laws" that will carry us throughout the book.

Let's take a deep sigh of relief.

EXERCISES

Solved Exercise 1.1: Show that the minimal model has steady-state glucose that depends on insulin sensitivity s and all other model parameters
Let's write the minimal model equations again,

(1.7)
$$\frac{dG}{dt} = m - sIG$$

(1.8)
$$\frac{dI}{dt} = qBf(G) - \gamma I$$

Steady state means no change with time, and thus we set the time derivatives to zero: $dG/dt = 0$ and $dI/dt = 0$. We find from Eq. 1.7, that

(1.9)
$$s\, I_{st} G_{st} = m_{st}$$

where m_{st} is the fasting production of glucose from the liver. The subscripts "st" denote steady state throughout this book.

Incidentally, this is the origin of the **HOMA-IR equation** used in research to estimate insulin sensitivity from steady-state glucose and insulin measurements:

$$s = m_{st}/I_{st}G_{st}$$

using the parameter $m_{st} = 22.5$ whose units assume that glucose is measured in mM and insulin in μU/mL (Matthews DR et. al., 1985).

Similarly, there is a useful equation for beta cell function, based on the steady state of Eq. 1.8, given by $qB = \gamma I/f(G)$. Often researchers use an approximation for $f(G)$, a straight line describing the slope near 5 mM glucose, $f(G) \sim G - 3.5$, providing the **HOMA-B equation** $qB = c\, I/(G - 3.5)$ with $c = 20$ in the units above.

To find the steady-state solution of the insulin equation, Eq. 1.8, let's approximate the regulation function $f(G)$ as $\left(\dfrac{G}{K}\right)^2$, as suggested by (Topp et al. 2000). This approximation is derived from the Hill function $f(G)=\dfrac{G^n}{K^n+G^n}$ with $n=2$ and is valid when $\left(\dfrac{G}{K}\right)^2 \ll 1$. This is not a terrible approximation since $\left(\dfrac{G}{K}\right)^2 \sim \left(\dfrac{5}{8}\right)^2 \sim 0.3$. Using this in Eq. 1.8, solved at steady state by setting $dI/dt=0$, we find that $q\,B\,(G_{st}/K)^2 = \gamma I_{st}$. Plugging this into Eq. 1.9, we obtain a steady-state glucose level G_{st} that depends on the cube root of all parameters (the cube root comes from the $\left(\dfrac{G}{K}\right)^2$ regulation):

$$(1.10) \qquad G_{st} = (\gamma K^2 \, m_{st}/s\; qB)^{1/3}$$

Let's consider the case of insulin resistance due to an 8-fold drop in s, keeping all other parameters the same. This results in a 2-fold rise in G_{st} because 2 is the cube root of 8. This is a rise from 5 to 10 mM, way past the criterion for diabetes. We see that G_{st} is not **robust** to changes in insulin resistance, which means it is sensitive to changes in this parameter.

We also see from Eq. 1.10 that glucose steady state is not robust to any of the other model parameters, including q, the maximal insulin production rate per beta cell. This parameter also changes because beta cell metabolism depends on many factors such as time of day (circadian rhythm), inflammation, and age. The parameter q also depends on **total blood volume**, as mentioned above, which dilutes the number of insulin molecules to units of concentration. Blood volume, which is about 5 L in adults, increases by 50% in pregnancy. It rises during childhood growth, in chronic exercise and in other physiological conditions. So being robust to q is also biologically important to achieve strict 5 mM control.

Note that when s becomes very small, the glucose nullcline crosses the insulin nullcline line when it saturates, that is when it flattens out. The expression becomes simpler. Because in this regime $f(G)=1$, we have at the fixed-point $\dfrac{m}{sG}=\dfrac{qBf(G)}{\gamma}=qB/\gamma$. Thus

$$G_{st} = m_{st}\,\gamma/sqB$$

The predicted rise in G_{st} when insulin resistance is large is dramatic, $G_{st} \sim 1/s$, emphasizing the inability of this simple feedback loop to explain normal glucose in people with obesity.

Solved Exercise 1.2: Show that half-life of glucose in the blood is not robust to insulin sensitivity in the minimal model

Likewise, the half-life of glucose in the blood is not robust. To see this, let's recall the removal term of glucose, namely $s\,I\,G$. The removal parameter – the factor multiplying G that has units of 1/time – is $a = s\,I$. The half-life, as discussed in the beginning of the chapter, is therefore $t_{1/2} = \ln(2)/a = \ln(2)/s\,I$. Let's consider the case that the system is at steady state, and we then add a small amount of glucose to the blood that hardly affects insulin concentration. Since at steady state $I = I_{st}$, the half life of the added glucose is $\ln(2)/s\ I_{st}$. We can compute I_{st} from Eqs. 1.8 and 1.9: $I_{st} = (m_{st}^2 qB/s^2\ \gamma K^2)^{1/3}$, and thus the half-life is $t_{1/2} = \ln(2)/s\ I_{st} = \ln(2)/(s\ m_{st}^2 qB/\gamma K^2)^{1/3}$. Therefore, glucose half-life depends inversely on the cube root of insulin sensitivity, $s^{-1/3}$. Half-life doubles if s shrinks by a factor of 8. Instead of returning to baseline within an hour of a meal, G stays high for 2 hours, surpassing criteria for diabetes. This is at odds with observations in most insulin resistant people.

Exercise 1.3. Additional biological features of the glucose-insulin circuit and diabetes
The goal of this exercise is to expand your knowledge of the glucose circuit and acquaint you with a nice video resource used by medical students. Watch the 19-minute video from osmosis.com on diabetes. Diabetes mellitus (type 1, type 2) and diabetic ketoacidosis (DKA) https://www.youtube.com/watch?v=-B-RVybvffU

a. Choose one element of the glucose system or diabetes (except glucagon) that we did not cover in this chapter. Read about it and summarize its role in glucose control and/or diabetes in 100 words.

b. Read about the hormone glucagon. Describe its role in 100 words.

c. Speculate on why the body needs two opposing hormones, insulin and glucagon? (100 words)

Exercise 1.4. Brain uptake of glucose
The brain takes up glucose from the blood at an insulin-independent rate. Modify the insulin-glucose model with a term describing this effect.

a. Write a formula for the steady states of glucose and insulin.

b. Is the steady-state blood glucose level G_{st} affected by the brain's uptake rate?

c. Plot the nullclines in this case. Is there still a single fixed point?

Exercise 1.5. Circadian changes in beta cells
Cells have clocks that track the time of day called circadian clocks. Beta cells secrete more insulin for a given level of glucose during the day than during the night. This can be modeled as a change in the parameter q in the insulin-glucose model. Suppose that during the day $q = q_1$ and during the night $q = q_2$ with $q_1 > q_2$.

a. What does the model predict for steady-state glucose and insulin during the day and night? Use the phase plot.

b. Is it unhealthy to eat big late-night meals?

Exercise 1.6. Simulate the glucose-insulin circuit

Write a computer code to simulate the dynamics of the glucose insulin circuit, Eqs. 1.7 and 1.8 in Solved Exercise 1.1. Use the following parameters $q = B = s = \gamma = 1$, $m = 0.5$, and $f(G) = G^2/(1+G^2)$.

a. Show that the steady-state solution is $G_{st} = 1$, $I_{st} = 0.5$.

b. Simulate the glucose tolerance test by adding a glucose input term that rises and falls with time, representing glucose arriving to the circulation from the gut, namely $m(t) = 0.5 + e^{-(t-5)^2}$. This glucose input function peaks at $t = 5$. Start the simulation at time $t = 0$ at steady state. Plot glucose and insulin as a function of time.

c. Simulate insulin resistance by setting $s = 0.1$. Repeat the simulation of the glucose tolerance test.

Here is a simple algorithm to get you started. Suppose you wish to solve the equations $dx/dt = f(x)$ with initial condition $x(0) = 0$.

```
x(0)=0  (initial condition)
t(0)=0  (time=0)
N=1000  (total time steps)
dt=0.1  (0.1s intervals)
for i=1:N
        t(i)=t(i-1)+dt
        dx(i)=f(x(i-1)) dt
        x(i)=x(i-1)+dx(i)
end
```

This numerical solution method is known as Euler's explicit time advancement.

Solved Exercise 1.7. Why is the rate of change equal to production minus removal?

Here is an explanation of the equations in this book in which the rate of change equals production rate minus removal rate. Suppose there are $X(t)$ molecules in the body at time t. Suppose that molecules are produced at rate p per unit time and removed at rate r per unit time. In a small time increment Δt, $p\Delta t$ molecules will be added and $r\Delta t$ will be removed, for a total of $(p-r)\Delta t$. Therefore, at time $t+\Delta t$, the number of molecules is $X(t+\Delta t) = X(t) + (p-r)\Delta t$. Rearranging this we find $(X(t+\Delta t) - x(t))/\Delta t = p - r$. If Δt is very small, the term on the left is just the differential dX/dt. We conclude that the rate of change of X is $dX/dt = p - r = $ production rate $-$ removal rate.

Solved Exercise 1.8. Why is the Hill equation justified biologically?

Here is a simple case in which the Hill equation applies. Suppose that a receptor is activated only when it binds n molecules of ligand L. The receptor can either be unbound and inactive, denoted R, or bound to n molecules at once and active, denoted RLn. The equilibrium constant for this reaction, $R+nL \rightleftarrows RLn$, is $Kd = [R][L]^n/[RLn]$, where $[R]$ is the unbound receptor concentration, $[RLn]$ is the bound receptor concentration and $[L]$ *is* the concentration of the ligand. Now since the receptor can only be in one of the two states, bound or unbound, we have $[R]+[RLn]=[Rt]$ where $[Rt]$ is the total receptor concentration. The fraction of active receptors is $f(L)=[RLn]/Rt=[RLn]/(R+[RLn])$. Using the expression for Kd, we find that the fraction of active receptors is $f(L)=\frac{L^n}{K^n+L^n}$, where the half-way point K is defined by $K^n=K_d$. Thus $f(L)$ has the form of the Hill equation. The Hill equation is a good approximation also in much more complex situations (Phillips, et al. 2012).

NOTE

1 Some additional features of the system are ignored here because they do not affect the core behavior of the feedback loop. This includes the following: Insulin secretion q is amplified by hormones released from the gut such as GLP-1 that sense incoming meals, and from brain inputs that anticipate meals. Insulin is secreted in two pulses, a brief spike of a few minutes followed by a prolonged insulin response to a meal, by beta cells that are heterogeneous with different sizes and secretion rates.

REFERENCES

Handbook of Diabetes, 4th edition, Rudy Bilous, Richard Donnely, 2010 ISBN: 978-1-444-39618-8 Wiley-Blackwell.

Alcazar, Oscar, and Peter Buchwald. 2019. "Concentration-Dependency and Time Profile of Insulin Secretion: Dynamic Perifusion Studies With Human and Murine Islets." *Frontiers in Endocrinology* 10. https://www.frontiersin.org/articles/10.3389/fendo.2019.00680.

Bergman, Richard N, Y. Ziya Ider, Y. Ziya Ider, Charles R. Bowden, Charles R. Bowden, Claudio Cobelli, and Claudio Cobelli. 1979. "Quantitative Estimation of Insulin Sensitivity." *The American Journal of Physiology*. https://doi.org/10.1172/JCI112886.

Bergman, Richard N. 2021. "Origins and History of the Minimal Model of Glucose Regulation." *Frontiers in Endocrinology* 11. https://www.frontiersin.org/articles/10.3389/fendo.2020.583016.

DeFronzo, Ralph A., Ele Ferrannini, Leif Groop, Robert R. Henry, William H. Herman, Jens Juul Holst, Frank B. Hu, et al. 2015. "Type 2 Diabetes Mellitus." *Nature Reviews Disease Primers* 1 (1): 1–22. https://doi.org/10.1038/nrdp.2015.19.

James, David E., Jacqueline Stöckli, and Morris J. Birnbaum. 2021. "The Aetiology and Molecular Landscape of Insulin Resistance." *Nature Reviews Molecular Cell Biology* 22 (11): 751–71. https://doi.org/10.1038/s41580-021-00390-6.

Kahn, S. E., R. L. Prigeon, D. K. McCulloch, E. J. Boyko, R. N. Bergman, M. W. Schwartz, J. L. Neifing, et al. 1993. "Quantification of the Relationship between Insulin Sensitivity and β- Cell Function in Human Subjects: Evidence for a Hyperbolic Function." *Diabetes*. https://doi.org/10.2337/diabetes.42.11.1663.

Kotas, Maya E., and Ruslan Medzhitov. 2015. "Homeostasis, Inflammation, and Disease Susceptibility." *Cell* 160 (5): 816–27. https://doi.org/10.1016/j.cell.2015.02.010.

Matthews DR, Hosker JP, Rudenski AS, Naylor BA, Treacher DF, Turner RC (1985). "Homeostasis model assessment: insulin resistance and beta-cell function from fasting plasma glucose and insulin concentrations in man". *Diabetologia*. **28** (7): 412–9. doi: 10.1007/BF00280883. PMID 3899825.

Osmosis video on Diabetes Mellitus: https://www.youtube.com/watch?v=-B-RVybvffU

Phillips, Rob, Jane Kondev, Julie Theriot, and Hernan Garcia. 2012. *Physical Biology of the Cell*. 2nd ed. Garland Science. https://doi.org/10.1201/9781134111589.

Strogatz, Steven H. 2018. *Nonlinear Dynamics and Chaos with Student Solutions Manual: With Applications to Physics, Biology, Chemistry, and Engineering*. CRC press.

Topp, Brian, Keith Promislow, Gerda Devries, Robert M. Miura, and Diane T. Finegood. 2000. "A Model of β-Cell Mass, Insulin, and Glucose Kinetics: Pathways to Diabetes." *Journal of Theoretical Biology*. https://doi.org/10.1006/jtbi.2000.2150.

CHAPTER 2

Dynamical Compensation, Mutant Resistance, and Type-2 Diabetes

W E NOW BUILD ON our work in Chapter 1 on the glucose-insulin system. We will understand how the system breaks down in type-2 diabetes and identify principles that apply broadly to other hormone circuits.

The insulin circuit is one of several dozen endocrine systems in the body. Endocrine organs secrete hormones that flow in the bloodstream and communicate with distant tissues. We will see that endocrine organs face three universal challenges. They must:

i. Work precisely even though they communicate with distant tissues that have unknown parameters that change over time. This is the problem of *robustness to parameter variations.*

ii. Maintain a proper organ size, even though cell populations tend to grow or shrink exponentially. This is the problem of *organ size control.*

iii. Avoid harmful mutant cells that can overgrow and take over the organ. This is the problem of *mutant resistance.*

We will discover a unifying and beautiful circuit design that addresses all three problems at once! This chapter also introduces the fundamental physiological laws that will accompany us through the book. Before we start, let's take a nice deep sigh of relief.

DOI: 10.1201/9781003356929-4

THE MINIMAL MODEL CANNOT EXPLAIN THE ROBUSTNESS OF GLUCOSE LEVELS TO VARIATIONS IN INSULIN SENSITIVITY

We ended the last chapter with a mystery. The insulin-glucose feedback loop of the minimal model explained the rise and fall of glucose after a meal but failed to explain how glucose levels are maintained when physiological parameters, like insulin sensitivity s, change.

The minimal model predicts that insulin resistance (low s) raises the glucose baseline above 5 mM and lengthens the response time in the glucose tolerance test. However, most people with insulin resistance, including people with obesity, maintain a normal 5 mM glucose steady-state concentration and exhibit normal glucose responses. The minimal model is thus not **robust** to parameters like s. It is also not robust to differences in blood volume, which dilute out insulin, or to the beta-cell maximal insulin production rate, q. In fact, the minimal model is not robust to *any* of its parameters.

Robustness must involve additional processes beyond the minimal model's glucose-insulin feedback loop. Indeed, the way that the body compensates for insulin resistance is by making more insulin. Each beta cell upregulates its insulin production capacity to the maximal possible. Then, there is an increase in the number and mass of beta cells. This is called beta cell *hyperplasia* – more cells – and *hypertrophy* – bigger cells. The two processes together increase the total mass of beta cells – with hypertrophy the dominant cause in humans after childhood. More beta cell mass means more insulin production. For example, people with obesity are insulin resistant and have more total beta cell mass than lean individuals. This extra secretion compensates for insulin resistance.

It's like factories making cars. To make more cars, one can increase production from each factory – but only up to a limit. Beyond that, more factories are needed.

Let's use the phase portrait to understand the effect of beta cell mass changes (Figures 2.1 and 2.2). The original set point of 5 mM glucose occurs at the intersection of the two nullclines. Insulin resistance shifts the blue nullcline and raises the fixed point to higher glucose (Figure 2.1). This is appropriate for short-term (hours to days) physiological changes in insulin sensitivity, such as acute stress or inflammation, where elevated glucose is useful. However, long-term excess glucose over weeks causes beta-cell mass to gradually increase. This raises the other nullcline, shifting glucose concentration back towards its original level (Figure 2.2). In this compensated state, insulin secretion is higher than in the original setpoint, due to the enlarged beta-cell functional mass.

For these shifts to produce precisely the right glucose level, 5 mM, beta cells must stop expanding at exactly the right mass (Figure 2.2). Remarkably, they do. **The resulting increase in insulin exactly compensates for the decrease in** s. Although each unit of insulin is less effective, the amount of insulin produced increases precisely enough to compensate.

This compensation is seen in a **hyperbolic relation**, in which healthy people show an inverse relationship between insulin sensitivity, s, and steady-state fasting insulin, I_{st}. This hyperbolic relationship, $I_{st} \sim 1/s$, maintains a constant product of the two variables: $s I_{st} = $ const (Kahn et al. 1993) (Figure 2.3). By contrast, for people with diabetes, the same product is lower (Figure 2.3, right). The origin of this hyperbolic relationship has long been a mystery; we will soon understand it.

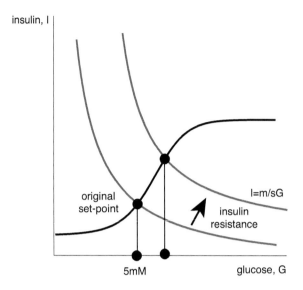

FIGURE 2.1   Insulin resistance raises the glucose set point.

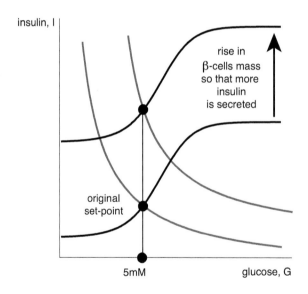

FIGURE 2.2   Beta-cell mass rises to restore the glucose set point.

A SLOW FEEDBACK LOOP ON BETA CELL MASS PROVIDES COMPENSATION

To explain how such precise compensation can come about, we need to extend the minimal model by adding an equation to describe how beta-cell total mass, B, can change.

Here we enter the realm of the **dynamics of cell populations**. These dynamics are unlike the dynamics of protein concentrations inside cells or molecules in the blood. For example, we used an equation for glucose that, at its core, has production and removal

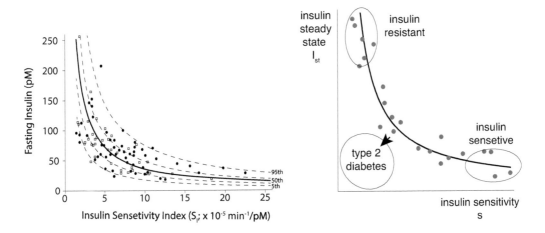

FIGURE 2.3 The hyperbolic relationship shows that fasting insulin is inversely proportional to insulin sensitivity in healthy people. Adapted from Kahn et al. (1993).

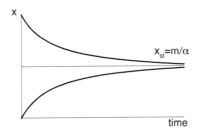

FIGURE 2.4 Circuits of molecular production and removal such as the glucose equation are inherently stable.

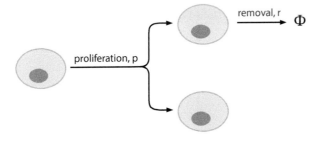

FIGURE 2.5 Cell circuits describe growth and removal of cell populations.

terms, $dG/dt = m - \alpha G$. Glucose safely converges to a stable fixed point, given by the ratio of production and removal rates $G_{st} = m/\alpha$ (Figure 2.4).

Cells, in contrast, **live on a knife's edge**. Their biology contains an inherent instability, due to exponential growth. Cells increase their biomass and divide (proliferate) at rate p and are removed at rate r (Figure 2.5). The removal rate includes active cell death (apoptosis)

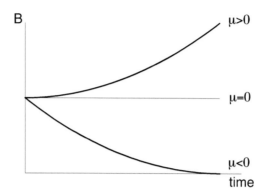

FIGURE 2.6   Cell circuits require stabilization because populations grow or decline exponentially unless net growth rate is precisely zero.

and other processes that take the cells out of the game like exhaustion, de-differentiation, and senescence. Since all cells come from cells, and all biomass is made by biomass, production of biomass is intrinsically autocatalytic. It is a rate constant p times the total mass of the cells: production $= p\,B$. Removal of beta-cell mass B is, as usual, B times the rate at which cells are removed: removal $= r\,B$. As a result, the change in total cell mass B is the difference between production and removal rates:

$$(2.1) \qquad dB/dt = pB - rB = (p - r)B = \mu B.$$

The key point is that the cell mass B appears in both growth and removal. It can therefore be taken outside the parentheses, leaving B times the **net growth rate** of cells, $\mu = p - r$, the difference between production and removal rates.

The problem is that if production exceeds removal, net growth rate μ is positive and total cell mass rises exponentially, $B \sim e^{\mu t}$ (Figure 2.6). Such explosive growth occurs in cancer. On the other hand, if removal exceeds production, net growth rate μ is negative, and cell numbers exponentially decay to zero, as in degenerative diseases. It is hard to keep total cell mass constant over time. This is known as the problem of **organ size control**.

Here we introduce the first of the three laws of physiology that are the foundation of this book:

Law 1: All Cells Come from Cells

The problem of organ size control is a natural outcome of this law.

Organ size control is an amazing and universal problem. Our body constantly replaces its cells; about **a million cells are made and removed every second**. We make and remove about 100 g of tissue every day (Sender and Milo 2021). If the production and removal rates were not precisely equal, we would exponentially explode or collapse.

To keep cell numbers constant, we need feedback control to **balance growth and removal** – to reach zero net growth rate, $\mu = 0$. Moreover, the feedback loop must keep the organ at a good functional size. Hence, the feedback mechanism must somehow register the biological activity of the cells and accordingly control their growth rate.

ORGAN SIZE CONTROL IN BETA CELLS IS PROVIDED BY FEEDBACK FROM GLUCOSE

Organ size control of beta cells is achieved by means of glucose, as pointed out by Brian Topp and Dianne Finegood (Topp et al. 2000). The feedback signal is blood glucose, which controls both the growth and removal rates. Thus $\mu = \mu(G)$. As measured in rodent islets, the removal rate of beta cells is high at low glucose and falls sharply around 5 mM glucose (Figure 2.7). Removal rate rises again at high glucose, a phenomenon called **glucotoxicity**, to which we will return soon.

For now, let's focus on the region around 5 mM. Biomass growth (which includes both cell division and growth of mass per cell) rises with glucose. For example, beta cells proliferate faster than normal in a fetus exposed to high glucose from a diabetic mother. Therefore, the curves describing the rates for growth and removal cross near $G_0 = 5$ mM, the fixed point that we seek with zero growth rate (Figure 2.8).

This way of plotting production and removal rates is called a **rate plot**, an important tool for understanding tissue-level circuits. The crossing point of the curves is the steady state, where cell production equals cell removal, and total cell mass does not change.

Another way of plotting this is to use the net growth rate μ, defined as the difference between production and removal. Net growth rate crosses zero at $\mu(G_0) = 0$ (Figure 2.9).

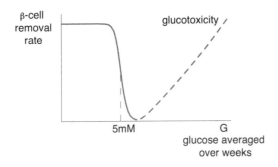

FIGURE 2.7 Beta-cell removal rate drops at about 5 mM glucose and rises again at high glucose levels.

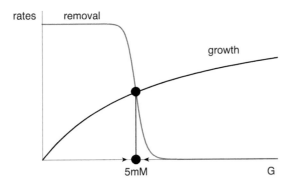

FIGURE 2.8 Beta cells reach a fixed point when growth and removal curves cross at about 5 mM.

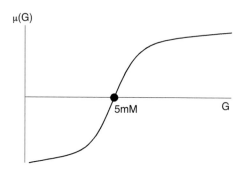

FIGURE 2.9   Beta cells reach a fixed point when their net growth rate equals zero.

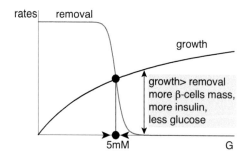

FIGURE 2.10   The 5 mM fixed point is stable because higher glucose makes beta-cell mass grow over weeks producing more insulin that pushes glucose back down; a similar restoring force occurs when glucose is lower than 5 mM.

The fixed point $G_0 = 5$ mM is **stable** for both beta cells and blood glucose. It is stable because perturbing glucose away from the fixed point causes it to move back. We can see this on our rate plot (Figure 2.10). If glucose is above 5 mM, beta cell mass grows faster than it is removed. Total beta cell mass increases, leading to more insulin, pushing glucose back down toward 5 mM. Conversely, if glucose is below 5 mM, beta cell mass is removed more rapidly than it grows, leading to less insulin, pushing glucose levels back up. These stable dynamics are indicated by the arrowheads pointing into the fixed point in Figure 2.10.

This feedback loop operates on the timescale of weeks, which is the growth rate of beta-cell biomass. It is much slower than the insulin-glucose feedback loop that operates over minutes to hours. The **slow feedback loop** of cell mass dynamics keeps beta cells at a proper functional mass and keeps glucose, averaged over weeks, at 5 mM.

The steep drop of the removal curve at $G_0 = 5$ mM is important for the precision of the glucose fixed-point. Due to the steepness of the removal curve, variations in growth rate (black curves) do not shift the 5 mM fixed point by much (Figure 2.11). The steep removal curve is generated by the cooperativity of enzymes that sense glucose and its byproducts inside beta cells (Karin et al. 2016).

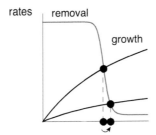

FIGURE 2.11   The glucose set point moves only slightly when beta-cell growth rate changes.

THE CELL MASS FEEDBACK CIRCUIT MAINTAINS HOMEOSTASIS DESPITE PARAMETER VARIATIONS

The slow feedback on beta cells can maintain a 5 mM glucose steady state despite variations in insulin sensitivity, *s*. To explain things in a quantitative way, we need to see this mathematically, not only graphically. To do so, let's add beta cell mass changes to the minimal model. This leads to a revised model, the BIG model which stands for the Beta-cell-Insulin-Glucose model, Figure 2.12. It is simply the two equations of the minimal model of Chapter 1 with a new equation, Eq. 2.4, for the total beta-cell mass *B*:

$$(2.2) \qquad dG/dt = m - s\,I\,G$$

$$(2.3) \qquad dI/dt = qBf(G) - \gamma I$$

$$(2.4) \qquad dB/dt = B\,\mu(G), \quad \mu(G_0) = 0$$

The only way to reach steady state in Eq. 2.4 is either at $B = 0$, meaning no beta cells at all, and therefore no insulin; or at zero net growth rate $\mu(G) = 0$, which occurs when $G = G_0 = 5\,\text{mM}$ glucose. The latter is the stable solution that describes healthy people. This powerful locking of glucose is similar to controllers in engineering known as integral feedback loops. If you want to know more about integral feedback in biology, see the Systems Biology course videos on my website or the book "Introduction to Systems Biology" (2019).

It is easy to calculate the steady state of the BIG model, thanks to Eq. 2.4, that locks glucose steady state at $G_{st} = G_0 = 5\,\text{mM}$. We can use this to find the insulin steady-state level by plugging G_{st} into Eq. 2.2, and setting $\dfrac{dG}{dt} = 0$, to find $I_{st} = m_{st} / sG_{st}$. The lower *s*, the higher

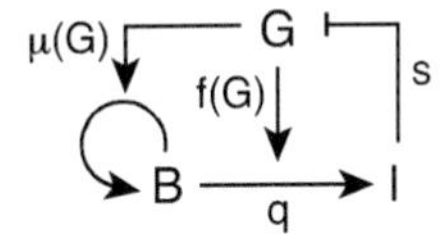

FIGURE 2.12   The BIG model circuit in which glucose controls beta-cell total mass, contributing a timescale of weeks to the faster insulin-glucose feedback loop.

the insulin concentration. This means that the product of insulin steady-state level and insulin sensitivity is constant,

$$sI_{st} = \frac{m_{st}}{G_{st}} = \text{const}$$

This explains the hyperbolic relation of Figure 2.3!

Finally, the beta-cell steady-state mass can be determined from Eq. 2.3, by setting $dI/dt = 0$, to find that

$$(2.5) \qquad\qquad B_{st} = \gamma \frac{I_{st}}{qf(G_{st})} = \frac{\gamma m_{st}}{sqG_{st}f(G_{st})}.$$

Interesting: beta-cell mass varies inversely with insulin sensitivity, $B \sim 1/s$. Beta-cell mass grows when s is small, as observed in people with insulin resistance. Beta-cell mass shrinks when insulin sensitivity is high, as in starvation. In fact, beta-cell mass varies with every parameter in the minimal model. Therefore, the organ-size control feedback makes **beta-cell mass expand or contract to precisely buffer out the effects of parameter changes**. It keeps the 5 mM steady-state despite variations in any of the minimal-model model parameters, including maximal insulin production per beta cell, q, insulin removal rate, γ, and even the fasting supply of glucose by the liver, m_{st}.

THE SAME CIRCUIT APPEARS IN MANY HORMONE SYSTEMS – A CIRCUIT MOTIF

The same circuit logic appears in many hormone systems that perform homeostasis, namely tight control of an important factor in the body. The cells that secrete a hormone in response to a signal also grow in response to the same signal.

For example, the concentration of free calcium ions in the blood is regulated tightly around 1 mM by a hormone called PTH, secreted by the parathyroid gland (Figure 2.13) (El-Samad, Goff, and Khammash 2002). The circuit has a negative feedback loop like insulin glucose, but with inverted signs: PTH causes increase of calcium, and calcium inhibits PTH secretion. The slow feedback loop occurs because parathyroid cell proliferation is regulated by calcium.

Other organ systems have similar "secrete and grow" circuits (Figure 2.14), in which the size of the organ expands or contracts to buffer variation in parameters. For example, thyroid hormone, essential for regulating metabolism, is secreted by the thyroid gland at

FIGURE 2.13 The secrete-and-grow circuit appears in the parathyroid hormone (PTH) system that controls blood calcium (Ca).

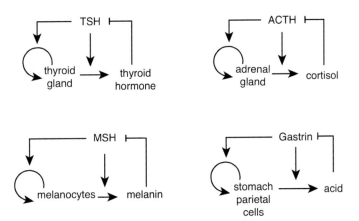

FIGURE 2.14 The secrete-and-grow circuit is a circuit motif because it appears in numerous hormonal systems.

the throat. The controlling signal is called TSH, which causes the thyroid gland cells to both secrete thyroid hormone and to proliferate. The thyroid is famous for overgrowing, sometimes to the size of a grapefruit, when more thyroid hormone is needed. This condition is called goiter.

Other systems with the same secrete-and-grow circuit are shown in Figure 2.14. They include acid secretion in the stomach by parietal cells under control of the hormone gastrin; secretion of cortisol by the adrenal gland under control of ACTH, and production of melanin by melanocytes in the skin under control of MSH. While the systems shown in Figure 2.14 differ in their molecular function, they share essentially the same circuit design as the insulin-glucose system. This is thus a **circuit motif.**

THE CIRCUIT ALSO MAKES THE DYNAMICS ROBUST

Remarkably, this circuit can also resolve the question of how glucose *dynamics* on the scale of hours are invariant to changes in insulin sensitivity. I mean that the BIG model shows how, in the glucose tolerance test, the response to a given input $m(t)$, such as drinking $75\,g$ of glucose, yields the same output curve $G(t)$, including the same amplitude and response time, for widely different values of the insulin sensitivity parameter.

This is unusual. Changing a key parameter in most models alters their dynamics. One might call such robustness of a dynamical response *rheostasis*, complementing the better-known concept of *homeostasis* which refers to maintaining a robust steady-state concentration of a metabolite.

This rheostatic ability was discovered by Omer Karin during his PhD with me (Karin et al. 2016). We named it **dynamic compensation** (DC): Starting from steady state, the output dynamics in response to an input is invariant with respect to the value of a parameter. To avoid trivial cases, the parameter must matter to the dynamics when the system is away from steady state (technically, to be observable). Solved Exercise 2.1 at the end of the chapter shows how dynamic compensation occurs in the BIG model. The proof is based on rescaling of the variables.

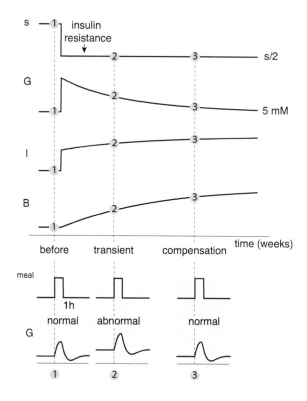

FIGURE 2.15 Dynamic compensation in the BIG model. A drop in insulin sensitivity causes glucose to rise within hours, but then beta-cell mass rises over weeks and glucose returns to baseline. The glucose response to a meal (bottom panel) is abnormal during the transient period of weeks but returns to normal when beta cell mass adapts to its new level.

Let's see how dynamic compensation works. We will use the **separation of timescales** in this system: cell mass changes much slower (weeks) than hormones (hours). Suppose that insulin sensitivity drops by a factor of 2, representing insulin resistance (Figure 2.15). As a result, insulin is less effective and glucose levels rise. Because glucose affects beta-cell growth rate, total beta cell mass rises over weeks (Figure 2.15 upper panels show the dynamics on the scale of weeks). More beta cells means that more insulin is secreted, gradually pushing glucose down to baseline. In the new steady state, there is twice the mass of beta cells and twice as much insulin. Glucose returns to its 5 mM baseline.

Let's now zoom in to the timescale of hours (Figure 2.15, lower panels). **The response of glucose to a meal before the drop in s is identical to the response long after the drop** (timepoint 1 and timepoint 3). In terms of glucose dynamics, the insulin resistance is invisible! The insulin response, however, is two times higher. Glucose dynamics in response to a meal are abnormal only during the transient period of days to weeks in which beta-cell mass has not yet reached its new, compensatory, steady state (timepoint 2).

Dynamic compensation thus allows people with different insulin sensitivity s to show the same glucose meal dynamics. Their insulin dynamics scale as $1/s$, namely more insulin when there is insulin resistance. This is indeed seen in experiments that follow non-diabetic

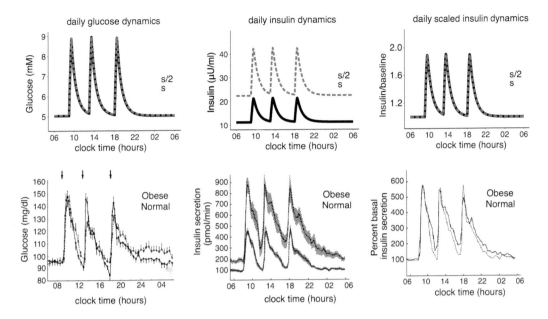

FIGURE 2.16 Non-diabetic obese people with insulin resistance show normal glucose responses to three meals. Their insulin is higher but resembles normal insulin when both are scaled to their baselines. Top panels show the BIG model (Karin et al. 2016), bottom panels show experimental data adapted from Polonsky et al. (1988).

people with and without insulin resistance over a day with three standardized meals (lower panels in Figure 2.16) (Polonsky, Given, and Van Cauter 1988). Their glucose levels rise and fall in the same way (lower left panel, Figure 2.16), but insulin levels are higher in people with insulin resistance (lower middle panel, Figure 2.16). As the model predicts, when normalized by the fasting insulin baseline, there is almost no difference in insulin between the two groups (lower right panel, Figure 2.16). The BIG model (upper panels in Figure 2.16) captures these observations.

This circuit seems so robust. What about diseases such as diabetes? How and why do things break down?

PREDIABETES IS DUE TO AN UPPER LIMIT TO BETA-CELL COMPENSATION

Before full-fledged diabetes sets in, there is a stage called **prediabetes** (Figure 2.17). In prediabetes, blood glucose shifts to higher and higher steady-state values, rising above 5 mM. Prediabetes is clinically defined by fasting glucose between 5.6 mM and the diabetes threshold of 6.9 mM. Prediabetes has no symptoms and occurs in 1 of 3 Americans, though 80% don't know that they have it. It is dangerous because people with prediabetes transit to type-2 diabetes at a rate of about 10% per year.

Prediabetes is often associated with insulin resistance. When insulin resistance is strong, beta cells must grow in functional mass by a large factor to compensate. But there is, in biology, always a limit to such compensation processes. This is our second law:

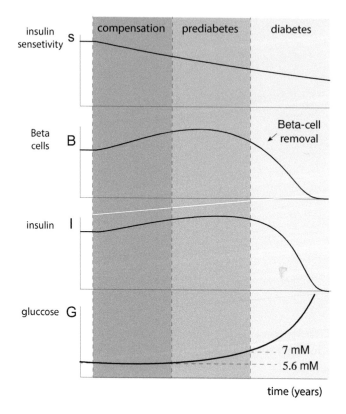

FIGURE 2.17 Prediabetes occurs when compensation for insulin resistance breaks down, due to a limit to beta cell mass growth, causing a rise in glucose. If glucose rise is unchecked, diabetes occurs and eventually beta-cell mass and insulin secretion drops.

Law 2: Biological Processes Saturate

In adulthood beta cells stop dividing. They can compensate by increasing their insulin secretion per unit biomass and the size of each beta cell. When functional beta-cell mass approaches its carrying capacity – determined by the maximal insulin secretion per unit biomass time the maximal possible size of a beta cell – compensation stops working. Beta cells hit a ceiling, and effectively the model returns to the minimal model of Chapter 1 with a constant beta-cell mass. Recall that the minimal model is not robust. Any further rise in insulin resistance causes glucose levels to rise above 5 mM. The stronger the insulin resistance, the higher the glucose.

One insight from the model is that prediabetes can result not only from low s but also from other parameter changes. As seen in Eq. 2.5, beta-cell mass goes as a specific combination of model parameters, $B \sim \dfrac{m\gamma}{qs}$. Thus, a decrease in beta-cell insulin production capacity q or insulin sensitivity s, or an increase in liver glucose production m or insulin removal rate γ, or a combination of these changes, can cause beta cells to hit their carrying capacity and compensation to saturate.

Having such a "parameter group" simplifies the understanding for the onset of disease. It also points to the way drugs or interventions work; for example, the diabetes drug metformin lowers liver production of glucose and thus lowers m, whereas exercise raises s. Both interventions prevent prediabetes but in different ways.

Another pathway to diabetes is a rapid rise in insulin resistance that is too fast for beta cells to grow and catch up. This happens in some cases in pregnancy, when insulin resistance rises due to signals secreted from the placenta in order to direct glucose toward the fetus rather than mom's cells. This is one cause of **gestational diabetes**.

Eventually, if untreated, prediabetes can lead to full-fledged type-2 diabetes. This disease shows a loss of beta cells and insulin, with a dramatic rise in glucose levels (Figure 2.17). At late stages, beta cells are gone, and the patient becomes dependent on insulin injections. We will next see that the transition to insulin-dependent type-2 diabetes is due to a dynamic instability that is built into the feedback loop.

TYPE-2 DIABETES IS LINKED WITH INSTABILITY DUE TO A U-SHAPED REMOVAL CURVE

Type-2 diabetes occurs when production of insulin does not meet the demand. Glucose levels go too high, damaging blood vessels.

The disease is linked with the phenomenon of glucotoxicity that we mentioned above: glucose at high levels disables and kills beta cells. Patients lose their beta cells and are not able to make enough insulin.

Glucotoxicity was quantified in an experiment by (Efanova et al. 1998) on rodent beta-cell islets. Islets were incubated for 40 hours in different concentrations of glucose. The fraction of dead islet cells dropped sharply at 5 mM glucose but then rose again above 10 mM glucose (Figure 2.18).

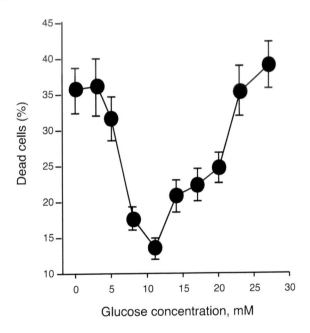

FIGURE 2.18 In glucotoxicity, beta cells are killed by high glucose levels as measured in rodent islets. Adapted from Efanova et al. (1998).

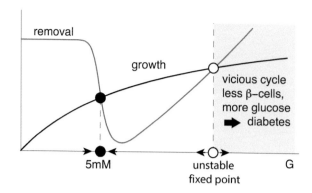

FIGURE 2.19 Glucotoxicity adds an unstable fixed point at high glucose levels, which leads to loss of beta cells when crossed.

The rate plot can help us see why glucotoxicity is so dangerous. It adds an **unstable fixed point**, the point at which proliferation rate crosses removal rate a second time (white circle in Figure 2.19). As long as glucose concentration lies below the unstable point, glucose safely returns to the stable 5 mM point. However, if glucose (averaged over weeks) crosses the unstable fixed point, beta-cell removal rate exceeds growth rate. Beta cells die, there is less insulin and hence glucose rises even more. This is a vicious cycle, in which glucose disables or kills the cells that control it. It resembles end-stage type-2 diabetes.

This rate plot can explain several risk factors for type-2 diabetes. The first risk factor is a diet high in fat and sugars. Such a diet makes it more likely that glucose fluctuates to high levels, crossing into the unstable region. A lean diet can move the system back into the stable region.

In fact, type-2 diabetes is often curable if addressed at early stages, by changing diet and exercising. This can bring average glucose G back into the stable region even if the unstable fixed point was crossed. G then flows back to normal 5 mM. Unfortunately, it is difficult for many people to stick with such lifestyle changes.

The second major risk factor is aging. With age, the growth rate of cells drops in all tissues including beta cells. This means that the unstable fixed point moves to lower levels of G (Figure 2.20), making it easier to cross into the unstable region. Note that the stable fixed point also creeps up slightly. Indeed, with age the glucose set point mildly increases in healthy people.

A final risk factor is genetics. A shifted glucotoxicity curve can make the unstable fixed point come closer to 5 mM (Figure 2.21).

Why does glucotoxicity occur? Much is known about *how* it occurs, which is different from *why* it occurs. Glucotoxicity is caused by programmed cell death that is regulated by the same processes that control beta cell growth and insulin secretion – glycolysis, ATP production and calcium influx. A contributing factor for cell death is reactive oxygen species (ROS) generated by the accelerated glycolysis in beta-cells presented with high glucose. Beta cells seem designed to die at high glucose – they are among the cells most sensitive to ROS, lacking the protective mechanisms found in other cell types.

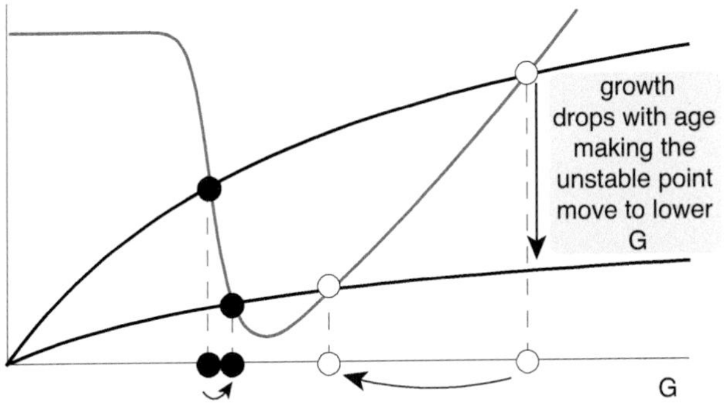

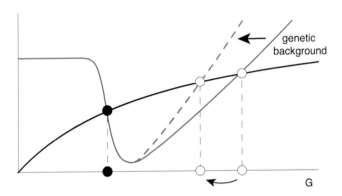

FIGURE 2.20 With age beta cell growth rate drops, bringing the unstable fixed point dangerously closer to the normal glucose set point.

FIGURE 2.21 Factors which shift glucotoxicity also shift the unstable fixed point.

Thus, it is intriguing to find a functional explanation for glucotoxicity – why is this dangerous effect not removed by natural selection?

Mainstream views are that glucotoxicity is a mistake or accident, exposed perhaps only recently due to our lifestyle and longevity. In this book, we take the point of view that such processes have an important physiological role. They are crucial in the young reproductive organism. Their benefit outweighs the cost of diseases in the old.

THE CIRCUIT IS FRAGILE TO INVASION BY MUTANTS THAT MISREAD THE SIGNAL

In 2017, Karin and Alon (2017) provided an explanation for glucotoxicity by considering a fundamental fragility of the organ size-control circuit motif. The fragility is to **take-over by mutant cells that misread the input signal**. Mutant cells arise when dividing cells make errors in DNA replication, leading to mutations. Mutations also arise passively over time, even in non-dividing cells. Rarely but surely, given the number of beta cells and the number of cell divisions in a lifetime,[1] a mutation will arise that affects the way that the cell reads the input signal. This is our third and final law:

Law 3: Cells Mutate

Let's examine such a mutation in beta cells. Beta cells sense glucose by breaking it down in a process called glycolysis, leading to ATP production, which activates insulin release through a cascade of events.

The first step in glycolysis is to modify glucose chemically. This is done by the enzyme glucokinase. Most cell types express a glucokinase variant that binds even tiny (micromolar) amounts of glucose, with a halfway-binding constant to glucose of $K = 40$ µM. But beta cells express a special variant with $K = 8$ mM. This half-way point is perfect for sensing the 5 mM range of glucose in normal conditions.

A mutation that affects the binding constant K of glucokinase, reducing it, say, by a factor of five, causes the mutant beta cell to mis-sense glucose concentration as if it were five times higher than it really is. The mutant beta cell therefore does glycolysis as if there was much more glucose around. It's as if the mutant "thinks" that glucose concentration G is actually 5G.

If our feedback design did not include glucotoxicity, such a mutant cell that interprets 5 mM glucose as 25 mM would have a higher proliferation rate (black curve) than removal rate (blue curve). It would think "Oh, we need more insulin!" and proliferate (Figure 2.22). The mutant cell therefore has a growth advantage over other beta cells, which sense 5 mM correctly. The mutant cell will multiply exponentially, particularly if this mutation occurs during embryonic development or early childhood when beta cells proliferate rapidly. This will eventually produce a substantial population of mutant beta cells. This is dangerous because such a population of mutant cells produces a lot of insulin, attempting to push glucose down to a set-point level that they think is 5 mM, but in reality is 1 mM, causing lethally low glucose.

Mutant expansion has a second, devious property: as the mutant cell population starts to push glucose below 5 mM, normal cells begin to be removed because their removal exceeds proliferation (they die to try to reduce insulin and increase glucose). The mutant's advantage is enhanced by killing off the normal cells.

Thus, biology has a challenge not usually seen in engineering. Suppose you want to control temperature; you use a thermostat. You can count on the thermostat being precise. It will not start dividing and mutating. Biology, in contrast, needs special designs to prevent takeover by mutant cells.

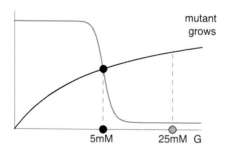

FIGURE 2.22 Without glucotoxicity, a mutant beta cell which misreads 5 mM glucose as 25 mM would grow.

BIPHASIC (U-SHAPED) RESPONSE CURVES CAN PROTECT AGAINST MUTANT TAKEOVER

To resist such mutant cells, we must give them a growth disadvantage. This is what gluco-toxicity does. The mutant cell misreads glucose as very high. As a result, its removal rate exceeds proliferation. The mutant kills itself (Figure 2.23). Mutants are removed. Isn't that neat?

The downside of this strategy is that it creates an unstable fixed point, with its vicious cycle. There is thus a **tradeoff between resisting mutants and resisting disease**.

In our evolutionary past, nutrition and activity probably prevented average glucose from being very high for weeks. The unstable fixed point was rarely crossed. Our modern lifestyle makes it more likely for glucose to exceed the unstable point, exposing a fragility to disease.

Glucotoxicity is a cell-autonomous strategy that eliminates mutants that strongly misread glucose. However, this strategy is still vulnerable to certain mutations of smaller effect – mutant cells that misread 5 mM glucose as a slightly higher level that lies between the two fixed points (hatched region in Figure 2.23). Such mutant cells still have a growth advantage, because they are too weak to be killed by glucotoxicity, and have higher proliferation rate than removal rate.

Designs that can help against intermediate mutants are found in this system: beta cells are arranged in the pancreas in isolated islets of ~1000 cells. A mutant might take over one islet, but not the entire organ. Relatively slow growth rates for beta cells also help keep such mutants in check. And, as we will see in Chapter 4, there are additional safeguards against these mutants, whose failure provides a mechanism for why the immune system attacks beta cells in type-1 diabetes.

The glucotoxicity mutant-resistance mechanism can be generalized to other organs: to resist mutant takeover of a tissue-level feedback loop, the feedback signal must be toxic at both low and high levels. Such U-shaped phenomena are known as **biphasic responses**, because their curves have a rising and falling phase. Biphasic responses occur across physiology. Examples include neurotoxicity, in which both under-excited and over-excited

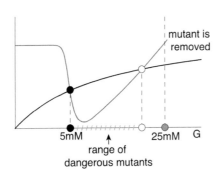

FIGURE 2.23 Glucotoxicity causes mutant beta cells that strongly hyper-sense glucose to kill themselves. This strategy has a loophole because mutant beta cells that only mildly hyper-sense glucose, in the range indicated by the hatched region, can still expand.

neurons die, and immune-cell toxicity at very low and very high antigen levels. These toxicity phenomena are linked with diseases, for example Parkinson's disease in the case of neurons.

Not all endocrine systems have such biphasic responses, however, requiring other mutant resistance strategies as we will study in Chapter 4.

SUMMARY

By modeling the glucose regulation system, we came upon new questions that reveal challenges shared by many organ-level circuits. First, organs have a fundamental instability due to exponential cell growth dynamics. They therefore require feedback to maintain a proper size. This is the problem of *organ size control*.

The feedback loops use a signal related to the tissue function – blood glucose in the case of beta cells – to make organ size and function arrive at a proper stable fixed-point.

A second fact of life for hormone circuits is that they operate on distant target tissues by secreting hormones into the bloodstream. The challenge is that the target tissues have variation in their parameters, such as insulin resistance. Hormone circuits thus need to be robust to such distant parameters in order to maintain good steady-state values (*homeostasis*) and dynamic responses (*rheostasis*) of the metabolites they control. We saw how hormone circuits can achieve this robustness by means of dynamic compensation. In dynamic compensation, tissue size grows and shrinks to precisely buffer the variation in parameters. As shown in the solved exercise below, dynamic compensation arises due to a symmetry of the equations.

Finally, organ-level feedback loops need to be protected from the unavoidable production of mutant cells that misread the signal and can take over the tissue. This problem of *mutant resistance* leads to a third principle: biphasic responses found across physiological systems, in which the signal is toxic at both high and low levels. Biphasic responses protect against strong mis-sensing mutant cells by giving them a growth disadvantage. This comes at the cost of fragility to dynamic instability and disease.

Thus, all three constraints – organ size control, robustness, and mutant-resistance – are addressed by a single integrated circuit design – the secrete-and-grow circuit. This circuit design is also found in numerous other hormone circuits.

EXERCISES

Solved Exercise 2.1: Show that the BIG model has dynamic compensation (DC)

To establish DC, we need to show that when starting at steady state, glucose output $G(t)$ in response to a given input $m(t)$ is the same regardless of the value of s. To do so, we will derive **scaled equations** that do not depend on s. To get rid of s in the equations, we rescale insulin to $\tilde{I} = sI$, and beta-cell mass to $\tilde{B} = sB$. Hence s vanishes from the glucose equation

(2.6) $$dG/dt = m - \tilde{I}G$$

Multiplying the insulin and beta-cell equations (Eqs. 2.5 and 2.6) by s leads to scaled equations with no s

(2.7)
$$\frac{d\tilde{I}}{dt} = q\,\tilde{B}f(G) - \gamma\tilde{I}$$

(2.8)
$$\frac{d\tilde{B}}{dt} = \tilde{B}\mu(G) \text{ with } \mu(G_o) = 0$$

Now that none of the equations depends on s, we only need to show that the initial conditions of these scaled equations also do not depend on s. If both the equations and initial conditions are independent of s, so is the entire dynamics.

There are three initial condition values that we need to check, for G, $\tilde{I}$, and $\tilde{B}$, which we assume begin at steady state at time $t = 0$. Note that if the system begins away from steady state, there is no DC generally. The first initial condition, $G(t = 0) = G_{st}$, is independent on s because $G_{st} = G_o$ is the only way for $\tilde{B}$ to be at steady state in Eq. 2.9. This means that the second initial condition, from Eq. 2.6, $\tilde{I}_{st} = m_0 / G_0$ is independent of s, which we can use in Eq. 2.7 to find that the third initial condition $B_{st} = \gamma\,\tilde{I}_{st} / G_0 f(G_o)$ is also independent of s. Because the dynamic equations and initial conditions do not depend on s, the output $G(t)$ for any input $m(t)$ is invariant to s, and we have DC.

Although $G(t)$ is independent on s, insulin and beta cells do depend on it, as we can see by returning to original variables $B = \tilde{B}/s$ and $I = \tilde{I}/s$. The lower s, the higher the steady-state insulin, as well as beta-cell mass, which rises to precisely compensate for the decreases in s.

Similar considerations show that the model has DC with respect to the parameter q, the rate of insulin secretion per beta cell, and hence to the total blood volume. There is no DC, however, to the insulin removal rate parameter, γ.

Dynamic compensation arises from the structure of the equations: the parameter s cancels out due to the linearity of the dB/dt equation with B, which is a natural consequence of cells arising from cells. s also cancels out from the dI/dt equation because the insulin production term, $q\,B\,f(G)$, is also linear in B, a natural outcome of the fact that beta cells secrete insulin.

Exercise 2.2. Brain uptake of glucose, BIG model

The brain takes up glucose from the blood at an insulin-independent rate.

a. Write a BIG model with a term describing this effect.

b. Write a formula for the steady states of glucose, insulin, and beta cells, G_{st}, I_{st} and B_{st}.

c. Is the steady-state blood glucose level G_{st} affected by the brain's uptake rate? Compare this to the minimal model.

d. Discuss why the BIG model design might be biologically useful when organs like the brain have varying fuel demands (50 words).

Exercise 2.3. The BIG model – numerical simulation
Write a computer code to numerically solve the BIG model equations. Set all parameter values to 1, $f(G) = G^2$ and beta-cell growth rate $dB/dt = 0.01 (G - 5)$. Note that due to the "0.01," the rate of change of $B(t)$ is much slower than the rate of change of $G(t)$ and $I(t)$. This represents the slow rate of beta-cell turnover compared to the fast hormone reactions.

a. Plot $G(t)$, $B(t)$, and $I(t)$ when at time $t = 100$, there is a drop of insulin sensitivity from $s = 1$ to $s = 0.2$. The plot should show the transition of $B(t)$ from one steady state to a new one. (Hint: The initial steady state of B is determined by setting all the time derivatives in the BIG model to zero.) Explain in 50 words.

b. Plot $G(t)$ and $I(t)$ in response to a meal, in the situation of (a). Model a meal by a pulse of glucose input. Thus, $m(t)$ goes from an initial value $m_0 = 1$ to a higher value $m_1 = 2$ for 1 time unit then back down to m_0. Let the meal begin at three different times, before, right after, and long after the drop of insulin sensitivity: $t_{meal} = 90$, 110, and 300. Plot a comparison of the response in the three meals in terms of how high and how quickly glucose rises and falls. Make sure the plots zoom in around the region of interest where glucose changes. Interpret using the concept of dynamical compensation (100 words).

Exercise 2.4. A model for prediabetes
In this exercise, we add to the BIG model a carrying capacity to beta cells and study the consequences of their loss of ability to compensate for parameter changes. The BIG model with carrying capacity is:

$$\frac{dG}{dt} = m - s\,I\,G$$

$$\frac{dI}{dt} = qBf(G) - \gamma\,I$$

$$\frac{dB}{dt} = B\left(p(G)\left(1 - \frac{B}{C}\right) - r(G) \right)$$

where $p(G)$ is beta-cell biomass growth rate, $r(G)$ is biomass removal rate, and the carrying capacity is C. The shapes of $p(G)$ and $r(G)$ are given schematically in Figure 2.19.

a. Since insulin has the fastest removal time, assume that I is at steady state and tracks the slower changes in B and G. Write the equation for I_{st} as a function of B and G.

b. Plug I_{st} into the other two equations. This reduces the model from three to two differential equations. Sketch the two nullclines for B and G. Note that the B nullcline has a shape of an inverted U (Figure 2.24).

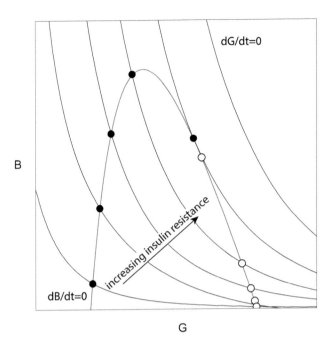

FIGURE 2.24 Phase plot of the slow variables B and G in a model with beta cell carrying capacity. The glucose steady state rises continuously with insulin resistance, until a critical value of insulin resistance where the stable and unstable fixed points collide and annihilate. Thereafter, beta-cell mass goes to zero and glucose rises.

c. How many fixed points are there? Interpret these fixed points in terms of healthy and diseased states.

d. Find steady-state glucose as a function of insulin sensitivity s. What happens to the 5 mM glucose set point when insulin resistance rises? What do changes in the other parameters do to the fixed points? Relate this to prediabetes.

e. When does the transition to late-stage type-2 diabetes occur, in which beta function is lost?

NOTE

1 Mutant cells that misread glucose are inevitable. Humans have about 10^9 beta cells. To generate, these cells required at least 10^9 cell divisions starting from the fertilized egg. The mutation rate is about 10^{-9}/base-pair/division. That means that every possible point mutation (single letter change in the genome) will be found in about 1 beta cell on average. Glucose mis-sensing can be caused by many different mutations, as exemplified by dominant activating glucokinase mutations (Christesen et al. 2008) so there should be multiple such mutant cells in everyone at birth. Human cells accumulate several tens of additional mutations per year even without dividing.

FURTHER READING

The BIG Model
Topp et al. (2000) "A model of β-cell mass, insulin, and glucose kinetics: Pathways to diabetes."

Dynamical Compensation
El-Samad et al. (2002) Calcium homeostasis and parturient hypocalcemia: an integral feedback perspective.
Karin et al. (2016) "Dynamical compensation in physiological circuits."

Resistance to Mis-sensing Mutants
Karin and Alon (2017) "Biphasic response as a mechanism against mutant takeover in tissue homeostasis circuits."

A General Resource for Models in Physiology
Keener and Sneyd, 2008 "Mathematical Physiology II: Systems Physiology"

REFERENCES

Christesen, Henrik B. T., Nicholas D. Tribble, Anders Molven, Juveria Siddiqui, Tone Sandal, Klaus Brusgaard, Sian Ellard, et al. 2008. "Activating Glucokinase (GCK) Mutations as a Cause of Medically Responsive Congenital Hyperinsulinism: Prevalence in Children and Characterisation of a Novel GCK Mutation." *European Journal of Endocrinology* 159 (1): 27–34. https://doi.org/10.1530/EJE-08-0203.

Efanova, Ioulia B., Sergei V. Zaitsev, Boris Zhivotovsky, Martin Köhler, Suad Efendić, Sten Orrenius, and Per Olof Berggren. 1998. "Glucose and Tolbutamide Induce Apoptosis in Pancreatic β-Cells: A Process Dependent on Intracellular Ca2+ Concentration." *Journal of Biological Chemistry* 273 (50): 33501–7. https://doi.org/10.1074/JBC.273.50.33501.

El-Samad, H., J. P. Goff, and M. Khammash. 2002. "Calcium Homeostasis and Parturient Hypocalcemia: An Integral Feedback Perspective." *Journal of Theoretical Biology* 214 (1): 17–29. https://doi.org/10.1006/JTBI.2001.2422.

Kahn, S. E., R. L. Prigeon, D. K. McCulloch, E. J. Boyko, R. N. Bergman, M. W. Schwartz, J. L. Neifing, et al. 1993. "Quantification of the Relationship between Insulin Sensitivity and β-Cell Function in Human Subjects: Evidence for a Hyperbolic Function." *Diabetes.* https://doi.org/10.2337/diabetes.42.11.1663.

Karin, Omer, and Uri Alon. 2017. "Biphasic Response as a Mechanism against Mutant Takeover in Tissue Homeostasis Circuits." *Molecular Systems Biology.* https://doi.org/10.15252/msb.20177599.

Karin, Omer, Avital Swisa, Benjamin Glaser, Yuval Dor, and Uri Alon. 2016. "Dynamical Compensation in Physiological Circuits." *Molecular Systems Biology* 12 (11): 886. https://doi.org/10.15252/msb.20167216.

Keener and Sneyd (2008) "Mathematical Physiology II: Systems Physiology."

Polonsky, K. S., B. D. Given, and E. Van Cauter. 1988. "Twenty-Four-Hour Profiles and Pulsatile Patterns of Insulin Secretion in Normal and Obese Subjects." *The Journal of Clinical Investigation* 81 (2): 442–48. https://doi.org/10.1172/JCI113339.

Sender, Ron, and Ron Milo. 2021. "The Distribution of Cellular Turnover in the Human Body." *Nature Medicine* 27 (1): 45–48. https://doi.org/10.1038/s41591-020-01182-9.

Topp, Brian, Keith Promislow, Gerda Devries, Robert M. Miura, and Diane T. Finegood. 2000. "A Model of β-Cell Mass, Insulin, and Glucose Kinetics: Pathways to Diabetes." *Journal of Theoretical Biology.* https://doi.org/10.1006/jtbi.2000.2150.

Three Laws of Physiology

I N CHAPTER 2, WE encountered three fundamental laws that we will use throughout the book:

1. All cells come from cells

2. Biological processes saturate

3. Cells mutate

Law 1 leads to exponential cell growth and the problem of organ size control. Law 2 limits compensation by organ size and opens a fragility to prediabetes. Law 3 requires mutant resistance strategies such as glucotoxicity, creating a fragility to type-2 diabetes.

To these three laws, we can add the force of natural selection. Physiology was selected to maximize transmission of genes to the next generation, not to optimize health or happiness. And since physiology was tuned by natural selection in the prehistoric environment, modern conditions can trigger mismatches that lead to disease.

We will see how these laws apply also to autoimmunity, aging, and age-related diseases in the upcoming sections of this book.

DOI: 10.1201/9781003356929-5

The Stress Hormone Axis as a Two-Gland Oscillator

INTRODUCTION

So far, we discussed an endocrine gland that controls its own size, with beta cells as our main example. We now explore what happens when *two* glands control each other's sizes. We will see that they create a feedback loop with a timescale of months, which can produce noise-driven oscillations. We will explore such feedback in the human stress-response pathway whose output is the hormone **cortisol**.

Cortisol is a hormone that prepares our body for physical and psychological stress. It is the main hormone in the stress-response circuit called **the HPA pathway**, involved in problems of chronic stress such as high blood pressure and heart disease, and in mood disorders such as major depression and bipolar disorder. Cortisol analogs are important drugs, used to suppress inflammation, and prevent tissue swelling. These drugs are gluco-corticoid steroids, often simply called **steroids**.

We will see how the textbook model for the HPA system works well for stresses on the timescale of minutes to hours. However, it cannot explain the aftermath of prolonged stress that lasts for months. For this longer timescale, we need to consider how the hormone glands change in size, affecting each other in a feedback loop. This will shed light on clinical phenomena on the scale of months, such as the side-effects of steroid withdrawal and the complex process of recovery from prolonged stress. The feedback loop also can act as a seasonal oscillator, providing a hormonal set-point for each season. Finally, we will explore this two-gland oscillator as a possible origin for the timescale of mood fluctuations in bipolar disorder.

THE HPA AXIS RESPONDS TO PHYSICAL AND PSYCHOLOGICAL STRESS

Let's begin with the classic description of the stress axis. When we wake up, we get a boost of energy (for some of us, including the author, this takes a bit of time), thanks in part to a morning surge of the hormone cortisol. We get a similar surge of cortisol if we see a threat,

DOI: 10.1201/9781003356929-6

such as a truck hurtling toward us. Cortisol increases blood pressure, releases sugar into the blood, focuses the mind, and gets us ready for fight or flight. We get a cortisol surge even if we *predict* that a threat is coming. This is psychological stress.

Cortisol takes minutes to act. Stresses also activate faster responses within seconds. These act through the sympathetic nervous system and secretion of adrenalin that makes our hearts beat fast and the liver secrete glucose. Without cortisol and adrenaline, we wouldn't have the energy to stand up.

The hormone circuit that controls cortisol is composed of a series of three glands, which we will denote H, P, and A, explaining its name, the *HPA* axis (Figure 3.1). The H gland is the hypothalamus, a brain region midway between our ears that receives neural inputs from other brain regions and uses this information to regulate hormones that act on the rest of the body. H is activated by emergencies and stresses, including pain, low glucose, low blood pressure, inflammation, and psychological stresses. It is also regulated by the brain's circadian clock which keeps track of the time of day, to generate the morning surge of cortisol and to keep cortisol low at night.

All these inputs are processed and combined into the output of H: secretion of the hormone CRH (corticotropin-releasing hormone) which we will call x_1. This hormone is secreted into a private blood vessel, called the portal system, which flows into the next gland P, a pea-sized organ at the bottom of the skull called the **pituitary**.

In P, the hormone x_1 stimulates cells to secrete the hormone x_2 (ACTH, adrenal cortex trophic hormone) into the circulation. Together with x_2, P also secretes beta-endorphin, which is a painkiller and causes euphoria. Endorphin means "endogenous morphine."

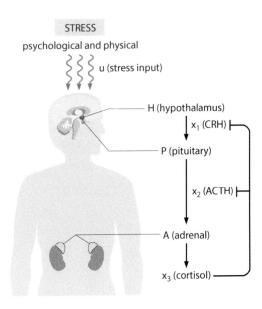

FIGURE 3.1 The HPA axis controls the production of the human stress hormone cortisol. In response to physical and psychological stressors the hypothalamus H secretes hormone x_1 (CRH) which induces the pituitary to secrete hormone x_2 (ACTH), along with beta-endorphins. x_2 induces the adrenal cortex to secrete cortisol. Cortisol inhibits the production of x_1 and x_2 in a feedback loop.

The messenger hormone x_2 flows with the circulation and reaches the third gland of the axis, A, the adrenal cortex. The word "cortex" refers to the outer layer of the adrenal gland.[1] We consider A as a single gland, although there are two adrenal glands, pyramid-shaped tips on top of the two kidneys. The cells in A secrete the final hormone, cortisol, denoted x_3. Cortisol closes a negative feedback loop by inhibiting the rate of production and secretion of its two upstream hormones, x_1 and x_2 (Figure 3.1).

Cortisol is a small fat-like molecule, a steroid hormone. Because it is fatty, it can penetrate the membranes of all the cells in the body and bind to cortisol receptors inside the cell. When these receptors bind cortisol, they attach to the DNA at specific sites, and cause the expression of genes that respond to stress. Cortisol has many effects in addition to increasing glucose and blood pressure, such as suppressing inflammation and affecting attention and memory. Cortisol diverts resources from projects like reproduction and growth toward immediate action. It gets the body ready to respond to the stress and to prepare for more stress.

Sudden stresses cause an **acute** response, in which cortisol levels rise from their normal range of 100–300 nM to nearly 1000 nM. These pulses last about 90 minutes, the half-life of cortisol. If the stress lasts for weeks or more, as in psychological anxiety, cortisol can cause the symptoms of **chronic stress**: weight gain, high blood pressure, risk of heart disease, diabetes, bone loss, and depression. Chronic high cortisol also causes cognitive changes, including increased sensitivity to negative stimuli and heightened anxiety, as well as decreased learning (technically, issues include decreased inhibition of the amygdala by the prefrontal cortex and damage to the hippocampus). The HPA axis is in fact implicated in mood disorders, as we will discuss later in this chapter.

That is one reason that when I teach a class, we take nice deep sighs of relief. Deep breaths relax the fast sympathetic nervous system for **fight and flight** and activate its countersystem called the parasympathetic nervous system for **rest and digest**. They cause a nice perception of the class as a pleasant place, which reduces stress. So let's all together take a nice deep sigh of relief.

Cortisol for optimal functioning needs to be in a middle range of concentration. Insufficient levels result in low energy and dangerously low blood pressure, whereas excessive levels over long times cause the symptoms of chronic stress (Figure 3.2). The midrange of inputs to the HPA axis can be thought of as healthy stimulation, as in sports and challenging activities that interest us.

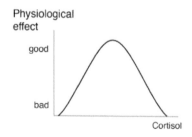

FIGURE 3.2 The physiological effect of cortisol is optimal at an intermediate level.

THE CLASSIC MODEL FOR THE HPA AXIS EXPLAINS RESPONSES ON THE TIMESCALE OF HOURS, BUT NOT ON THE TIMESCALE OF MONTHS

This is the description of the HPA axis that you find in textbooks. It's the HPA version of the minimal model we saw for glucose and insulin. Let's write it down mathematically so that we can explore its properties (Vinther, Andersen, and Ottesen 2011). The main conclusion will be that the model shows the expected response to acute stress over hours but cannot explain phenomena on the scale of weeks-months. We will then expand the model by adding control of gland sizes and explain a world of phenomena on the month timescale.

Let's lump together all the stress inputs from the brain into a single quantity $u(t)$. This input u causes H to secrete the hormone x_1, which is degraded with a timescale of minutes at rate α_1. This can be described as follows, using the same logic as the glucose minimal model of Chapter 1, namely production minus removal:

$$(3.1) \qquad \frac{dx_1}{dt} = q_1 \frac{H\,u}{x_3} - \alpha_1 x_1.$$

the inhibitory effect of x_3 on the secretion of x_1 is described by x_3 in the denominator of the production term[2] so that the more cortisol x_3, the lower the production rate of x_1.

The hormone x_1 acts on pituitary cells (called corticotrophs), whose total mass is P, to induce secretion of the second hormone x_2. Its secretion is also inhibited by x_3. The hormone x_2 has a lifetime of $1/\alpha_2 = 10$ minutes, and so

$$(3.2) \qquad \frac{dx_2}{dt} = q_2 \frac{P\,x_1}{x_3} - \alpha_2\,x_2$$

The hormone x_2 in turn induces the cells of the adrenal cortex A to secrete x_3, whose lifetime is $1/\alpha_3 = 90$ minutes:

$$(3.3) \qquad \frac{dx_3}{dt} = q_3\,A\,x_2 - \alpha_3 x_3$$

To solve for the steady-state hormone levels, we set all temporal derivatives to zero as in the previous chapters, and do the algebra. The cortisol steady state depends on all the model parameters and rises with the input u and with the sizes of the two glands A and P

$$(3.4) \qquad x_3^{st} = \left(\frac{q_1 q_2 q_3}{\alpha_1 \alpha_2 \alpha_3}\,u\,HPA \right)^{\frac{1}{3}} u^{\frac{1}{3}}\,P^{\frac{1}{3}}\,A^{\frac{1}{3}}$$

the 1/3 powers come from the three steps in the *HPA* axis. Similarly, the steady states of the other two hormones also depend on all parameters. x_2 rises with the size of P but drops with the size of A, because A inhibits its upstream hormone x_1

(3.5)
$$x_2^{st} \sim u^{\frac{1}{3}} A^{-\frac{2}{3}} P^{\frac{1}{3}}$$

(3.6)
$$x_1^{st} \sim u^{\frac{2}{3}} P^{-\frac{1}{3}} A^{-\frac{1}{3}}$$

where the $\sim$ sign signifies an expression that highlights the dependence on the variables we care about, without the other parameters. These equations correctly show that the hormone levels rise when stress input u increases and drop back when the stress is removed, within minutes-hours (Figure 3.3).

The equations, however, cannot explain phenomena on the scale of weeks-months mentioned in the introduction to this chapter. They predict that cortisol returns to normal within hours of the end of the stress input. This cannot explain the months required for hormone levels to return to baseline after prolonged HPA activation due to stress, pregnancy, anorexia, drug abuse, or long-term use of steroids in clinical situations.

To explain these phenomena, we will need the slow timescale of months associated with the growth and shrinkage of the glands in the HPA axis. The fact that the adrenal grows upon prolonged stress has been known for decades. Enlarged adrenals in stressed rats and in human suicide victims were one of the first clues for the existence of entire pathway as discovered by Hans Selye. But the dynamical consequences of these gland-mass changes have not been considered mathematically until recently.

We will therefore study the effect of adding equations for the gland masses. As in the case of beta cells, the glands A and P are made of cells that divide and are removed (the H gland is in the brain, and like most neuronal tissues, it does not show sizable turnover of cells). To maintain proper **organ size control**, the A and P glands must have feedback control. It turns out that the glands control each other's growth using their series of hormone secretions. This circuit is described next.

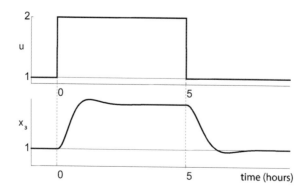

FIGURE 3.3 The HPA model in textbooks explains the rise of cortisol over hours in acute stress.

THE HPA AXIS SHOWS A FEEDBACK LOOP IN WHICH TWO GLANDS CONTROL EACH OTHER'S SIZE

The cells of the A and P glands proliferate and are removed with a typical turnover time of about 1–2 months. They balance their cell proliferation and removal with feedback loops. The inputs to these loops are the *HPA* hormones themselves. The hormone x_1 acts not only to induce secretion of x_2 by P cells; it also increases the P cell proliferation rate. Similarly, x_2 acts to induce x_3 secretion and also to induce adrenal cortex A cell proliferation (Figure 3.4).

Thus, **the HPA axis shows two occurrences of the secrete-and-grow circuit motif we saw for the insulin-glucose system**, in which a signal controls both secretion and growth (Figure 3.4). In the *HPA* axis, the two motifs are stacked on top of each other.

To reflect for a moment, we have learned quite a lot: before reading this book, it probably would have been difficult for you to make sense of such a complex circuit, but in about 10 minutes, you will understand much of its behavior.

The equations for P and A cell mass are, just as in the case of beta cells, the difference between cell proliferation and removal. The proliferation of P depends on x_1, and proliferation of A on x_2, and thus

$$(3.7) \qquad \frac{dP}{dt} = P\left(b_P x_1 - a_P\right)$$

$$(3.8) \qquad \frac{dA}{dt} = A\left(b_A x_2 - a_A\right)$$

where a_P, a_A are the cell removal rates (with turnover times of $1/a_A$, $1/a_P \sim 1-2$ months), and b_P, b_A are the hormone-dependent growth rates.

These two equations have important consequences for the system on the scale of months. Here we again use the concept of **separation of timescales** between the month-timescale of changes in gland mass and the hour-timescale of the hormones. Therefore, the steady states of Eqs. 3.4–3.6 correctly describe what happens if the stress input u is present for a few hours. But if stress input u is present for weeks, the masses of the glands A and P increase gradually, and the system finds its full steady state with an enlarged adrenal cortex and high cortisol (Figure 3.5). The fast-timescale solutions for the hormones in Eqs. 3.4–3.6 are more accurately described as **quasi-steady-states**.

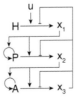

FIGURE 3.4 The HPA axis is built of two secrete-and-grow circuits in series. The hormone x_1 is the growth factor of the pituitary cells P that secrete x_2, the growth factor for the adrenal cortex A.

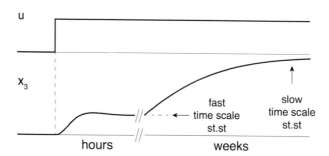

FIGURE 3.5 Separation of timescales in the HPA model with gland-mass changes. The *HPA* axis has a fast timescale of hours and a slow timescale of weeks over which gland masses change. A prolonged stress input causes the adrenal cortex to grow and cortisol levels to rise over the course of weeks toward an elevated steady-state level.

When we solve the steady-states of the slow equations Eqs. 3.7 and 3.8, we find that the steady-state hormone levels become much less sensitive to the physiological parameters of the *HPA* axis. In particular, the only way to get to a steady state (with A, $P>0$) is to have cell proliferation equal removal. This locks the hormone x_1 to obey $b_p x_1 = a_p$, and thus $x_{1,st} = a_p/b_p$. Similarly, $x_{2,st} = a_A/b_A$. Elegant! Only cell growth and removal parameters determine hormone levels, making them robust to all the fast-timescale parameters such as production rates q_1, q_2, q_3, removal rates α_1, α_2, α_3, etc.

Cortisol similarly turns out to have a simple steady-state level that is robust to almost all model parameters (by plugging in x_1^{st} to Eq. 3.1):

(3.9)
$$x_3^{st} = \frac{q_1 b_p H}{\alpha_1 a_p} u$$

Nicely, cortisol rises linearly with the stress input u – a proportional response to stress inputs from the brain. Cortisol steady-state concentration does not, however, depend on production parameters q_2, q_3 which vary, for example, with metabolism. It also does not depend on the hormone removal parameters α_2, α_3. As in the case of glucose, the circuit protects cortisol baseline levels from variations in many physiological parameters, compensating for these changes by means of slow growth and shrinkage of the glands.

The ability of gland masses to change also provides robustness to the response dynamics to an acute stress, the phenomenon called rheostasis. Cortisol rise and fall dynamics after an acute stress input are independent of the production rates q_2, q_3, an example of dynamical compensation that we saw in the previous chapter for glucose.

FIGURE 3.6 The adrenal and pituitary form a negative feedback loop that acts on the timescales of months.

We can derive equations for the gland masses on the timescale of months by using the quasi-steady-state approximation for the hormones. Plugging x_1 and x_2 into Eqs. 3.7 and 3.8, shows a **negative feedback loop between P and A:**

$$(3.10) \qquad \frac{dP}{dt} = P\left(c_P u^{\frac{2}{3}} P^{-\frac{1}{3}} A^{-\frac{1}{3}} - a_P \right)$$

$$(3.11) \qquad \frac{dA}{dt} = A\left(c_A u^{\frac{1}{3}} P^{\frac{1}{3}} A^{-\frac{2}{3}} - a_A \right)$$

where c_P and c_A are combinations of the fast-timescale equation parameters. P acts to increase the size of A, by the action of x_2 the growth factor for the adrenal cortex (Figure 3.6). In contrast A prevents the growth of P, because the hormone it secretes, cortisol, reduces the secretion of x_1, which is the growth factor for P.

THE PITUITARY AND ADRENAL GLANDS FORM A NEGATIVE FEEDBACK LOOP THAT CAN OSCILLATE WITH A SEASONAL TIMESCALE

For the remainder of this chapter, we focus on the timescale of months. This is the timescale of the negative feedback loop between the two glands. Like many negative feedback loops in which both arms have similar timescales, the *HPA* circuit can show damped oscillations.

To see how the oscillations go, follow Figure 3.7 from the top and go with the arrows clockwise. When A is large, it reduces P. Smaller P reduces A. Smaller A makes P grow back. Large P makes A grow, closing the cycle (Figure 3.7). Each of these steps has a delay due to the tissue turnover time, about 2 months. The overall time for a full cycle is **on the order of a year**.

When Avichai Tendler during his PhD with me found this timescale of a year, we got goosebumps (Tendler et al. 2021). A year is important for organisms because of the changes in seasons. It is useful to have **an internal oscillator that can keep track of seasons**, as we will discuss below.

To understand this oscillator, it helps to make an analogy of the A-P feedback loop to a mass on a spring. Suppose the mass is at rest at its equilibrium point (Figure 3.8). If you

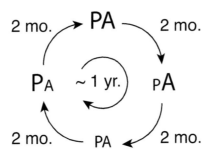

FIGURE 3.7 Schematic of the oscillatory nature of the adrenal-pituitary feedback loop. A large adrenal A secretes cortisol that inhibits x_1, the growth factor of the pituitary P, making it shrink. Small P produces less x_2, the growth factor for A, so that A shrinks. Small A releases the inhibition of P growth, and a large P makes A grow again. Each step takes months.

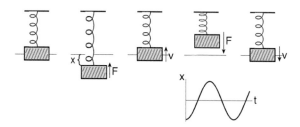

FIGURE 3.8   The adrenal-pituitary feedback loop can be compared to a mass on a spring.

FIGURE 3.9   In the mass-on-a-spring analogy there is a negative feedback loop between velocity v and position x.

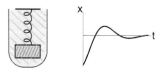

FIGURE 3.10   A better analogy for the adrenal-pituitary feedback loop is a mass on a spring in a viscous fluid, which exhibits damped oscillations.

pull the mass down, the spring will stretch and exert a force pointing back to the equilibrium point. If you let go of the mass, it will return to equilibrium but with high velocity v, so it will overshoot and compress the spring. The compressed spring pushes the mass back down, and so on, to obtain sustained oscillations (Figure 3.8).

To describe this motion, we can use Hooke's law of the spring: the force grows with the extension of the spring, $F = -kx$. This force, according to Newton's law, $F = ma$, changes the acceleration a, which is the rate of change of velocity $a = dv/dt$. Thus, spring extension x "inhibits" velocity, $dv/dt = -\dfrac{k}{m}x$. Conversely, velocity "activates" spring extension because velocity is the rate of change of position, $dx/dt = v$. This is a negative feedback loop in which v enhances x but x inhibits v (Figure 3.9). Thus, x and v are like A and P, and the negative feedback can act as an oscillator.

When you simulate the slow equations for A and P, you don't get sustained oscillations; instead, you see damped oscillations that decay away. To get damped oscillations in the spring analogy, we can submerge the spring in a container with a viscous fluid like honey. The honey causes a **drag** force which is proportional to velocity: $dv/dt = -kx - bv$ (Figure 3.10). The spring does not oscillate forever but instead settles down to its equilibrium point.

There is also a drag-like force in the equations for A and P. For example, large A makes more cortisol, which inhibits x_2, the growth factor of A. Thus, large A acts to reduce its own growth, similar to a large velocity v which enhances drag and slows itself down (thanks to Z. Tan in the 2020 Systems Medicine class for this comment).

DAMPED OSCILLATIONS IN STEROID WITHDRAWAL OVER MONTHS

When you pull the damped spring and let it go, it oscillates and relaxes to its steady state. This suggests a way to study the *HPA* oscillator, by seeing what happens when you pull it away from equilibrium and watch it recover. Let us therefore consider situations that perturb the *HPA* axis, reasoning that the two-gland feedback loop should show months-scale recovery dynamics from perturbations.

One situation occurs when cortisol levels are forced to be high for weeks or more, and then are suddenly lowered. This is a common medical situation in which people take cortisol analogues called glucocorticoid steroids, such as dexamethasone or prednisone, for extended periods and then go off the drugs (Dixon and Christy 1980; Byyny 2009). Millions of people take glucocorticoid steroids to reduce inflammation or suppress immune responses, as in asthma, autoimmune diseases, and after transplants.

High doses of steroids for a few days cause no problems. But if they are given for 2 weeks or more, it is important not to stop steroid treatment all at once. A sudden stop can make the patient show dangerously low cortisol. The adrenals can't make enough cortisol, causing serious symptoms: blood pressure drops to potentially fatal levels. This is called **steroid withdrawal**. Thus, one must gradually reduce the dose over months – one protocol is to reduce the dose by 25% every 2 weeks.

Steroid treatment can be modeled by adding external cortisol D to the equations, by replacing x_3 by $x_3 + D$ in Eqs. 3.1 and 3.2. In this case, both pituitary and adrenal gland sizes shrink, due to inhibition of their growth factors x_1 and x_2 by D. It's as if the glands "think" there is too much cortisol and shrink in an attempt to return it to baseline. This causes the atrophied and involuted adrenal gland observed in extended glucocorticoid treatment (Nicolaides et al. 2010).

We can use our mathematical model to simulate steroid withdrawal. The computer simulation begins with small gland sizes A and P, after a long period of high exogenous cortisol D. A sudden withdrawal from the drug is modeled by setting D to zero. This is like pulling hard on the spring, and then letting it go. You can see this in Figure 3.11, starting at the bottom (small A and P), and following the arrows. The dynamics show that P recovers first, and A follows it. As a result, x_2 overshoots after 3 months and goes back to normal after 9 months, whereas cortisol is abnormally low and recovers after 9 months (Figure 3.11a). The overshoot of x_2 is due to the release of x_1 inhibition, which causes P to grow. Only when P returns to normal size is x_2 sufficient to allow A to grow and recover.

Such overshoot dynamics were found by Graber et al. in an example of a "small data" study which followed $n = 14$ patients who went off prolonged steroid treatment or had a cortisol-secreting tumor removed (Graber et al. 1965). Patients showed a 10-fold overshoot in x_2 after a few months, and a slower recovery of cortisol over 10 months (Figure 3.11b), in agreement with the model.

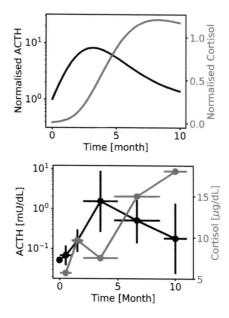

FIGURE 3.11   Cortisol and x_2 (ACTH) dynamics following sudden steroid withdrawal. (a) The model shows cortisol recovery after 9 months with an overshoot in x_2. (b) Data from 14 patients which stopped prolonged steroid treatment or had a cortisol-secreting tumor removed shows similar dynamics. Adapted from Graber et al. (1965).

We can further understand this overshoot by using the phase portrait approach we introduced in Chapters 1 and 2 (Figure 3.12). The two nullclines decompose the feedback loop of A and P into two arms. One arm ($dA/dt=0$) is a rising curve in which P enlarges A, and the other arm ($dP/dt=0$) is a dropping curve in which A reduces P. The arrows in the phase portrait show how the two glands "flow" back to their fixed point at the intersection of the nullclines. The flow spirals into the fixed point.

I love the look of this phase portrait. See how steroid withdrawal is easy to read out (blue trajectory) by following the arrows: first P grows, then A, and then a spiral back to the fixed point.

RECOVERY FROM PROLONGED STRESS TAKES WEEKS AND SHOWS AN ENDORPHIN UNDERSHOOT

Another situation that pulls hard on the spring is prolonged stress, namely an input u that is high for weeks or more (Karin et al. 2020). This occurs in periods of severe disease, starvation, psychological anxiety or depression. Prolonged activation also occurs in periods of alcoholism or drug abuse, because alcohol and many drugs activate the HPA axis (Karin, Raz, and Alon 2021).

Simulating the model with prolonged high input u shows that the adrenal mass A grows over weeks, and cortisol rises. Note that this is the opposite of steroid treatment in which the adrenal shrinks; here we increase the input u, rather than a steroid that mimics cortisol.

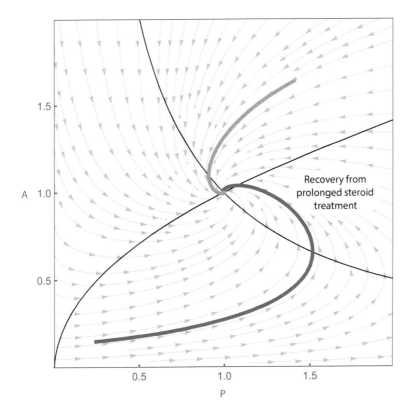

FIGURE 3.12 Phase portrait for the HPA axis gland masses. The blue trajectory corresponds to recovery from prolonged steroid treatment and shows a strong overshoot of the pituitary. The orange trajectory corresponds to recovery from prolonged stress and shows an undershoot of the pituitary. Black curves are nullclines that cross at the fixed point.

Unlike cortisol, the hormones x_1 and x_2 first rise during the stressful period but then return to their original level, despite the ongoing stress – recall the integral feedback equations Eqs. 3.7 and 3.8 that make their steady-state independent on input u. This is called exact adaptation – return to a baseline despite continued strong input.

Ok, we pulled the spring, now let's release it. Stress, illness or starvation is over; addiction stops cold turkey. Our phase portrait shows that it takes weeks for the adrenal to return to normal, and the pituitary shows an undershoot (Figure 3.12 orange trajectory that starts from an enlarged adrenal). As a result, it takes a month or so for cortisol to go back to normal. Importantly, there is a longer period of low x_2, which as you recall also means low levels of the natural painkiller and euphoric beta-endorphin.

Thus, the model reveals two periods of post-stress adaptation (Figure 3.13). There is high cortisol and low beta-endorphin for a few weeks, and then a few more months of normal cortisol in which beta-endorphins remain low (Karin, Raz, and Alon 2021). Low beta-endorphins can cause dysphoria – a reduced ability to enjoy pleasurable stimuli. And in some people, it can also increase sensitivity to pain. We need care and compassion in the months when people recover from prolonged stressors.

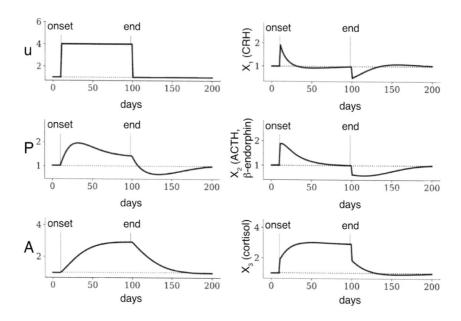

FIGURE 3.13 HPA dynamics during and after a prolonged stress input. The adrenal A and cortisol x_3 both grow during the stress and then decline when it ends. Hormones x_1 and x_2 shows a pulse-like overshoot at stress onset with exact adaptation, and an undershoot at stress end. There is a period of months in which x_2 and beta-endorphins remain low after cortisol has normalized. Adapted from Karin et al. (2021).

SEASONALITY IN HORMONES

With the mass-on-a-spring analogy, we can also understand what happens to the HPA axis when it gets inputs that change over a year, like the changes of the seasons. These can **entrain** the oscillator to a period of one year. It's like putting the spring-in-honey container on a platform that oscillates up and down: the spring picks up the platform's frequency (one year) and begins to oscillate with the same frequency. Even if the oscillator's natural frequency is a bit different, perhaps 10 months, it can still entrain effectively to a yearly input.

Many animals exhibit striking physiological changes across the year – they migrate, hibernate, and mate in certain seasons, all due to hormonal changes. The signals for this seasonal cycle are day-length variation, shortest in midwinter and longest in midsummer (Figure 3.14). Short days are a kind of stress input. Day length affects a brain hormone called melatonin, which feeds into the P gland.

We can thus ask about human hormone seasonality.

But day length signals are subtle, and one does not want to be fooled by a few cloudy days in summer into thinking it is winter. This requires a system that can integrate inputs over weeks and keep track of the seasons. The long timescale of the $A – P$ oscillator is well suited to provide the needed inertia.

To study the effects of entrainment by yearly cycles of day length, we can provide the HPA axis model with inputs that vary with a period of 1 year, $u(t)=1+u_0 \cos(\omega t)$.

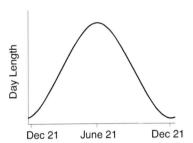

FIGURE 3.14   Seasonal behavior is governed by day length which is shortest on December 21.

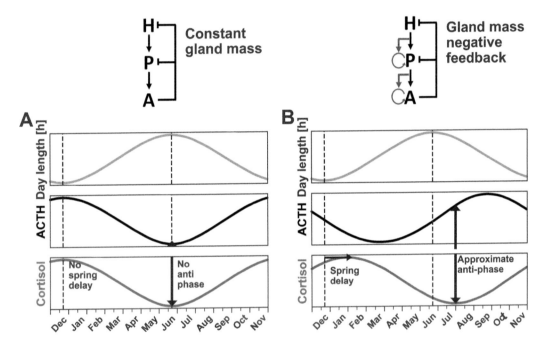

FIGURE 3.15   An HPA model with no gland-mass (A) changes predicts that hormones track the day length input with a maximum on December 21. A model with gland-mass (B) changes shows a delayed peak of cortisol in late winter and an antiphase between cortisol and its upstream hormone x_2. These phase shifts are due to the month-scale turnover of the glands.

A model with no gland-mass changes would predict that hormones simply track the day length input. When input is highest, on December 21, all hormones are highest.

Adding gland-mass changes, however, modifies this picture and provides specific predictions (Figure 3.15):

i. Peak cortisol should not occur at the time of peak input (December 21) but rather be delayed by about 2 months. This is a **spring-shift** in cortisol.

ii. Cortisol and x_2 (ACTH) are in approximate **antiphase**, with x_2 peaking in late summer/fall.

This spring-shift and antiphase cannot occur in a model without gland size changes, because in such models the hormones vary together with the input and with each other within minutes-hours. The spring shift and the antiphase can be computed precisely using linearized forms of the equations for the gland masses; these phases depend only on two parameters – the turnover times of the P and A glands.

LARGE-SCALE MEDICAL RECORDS SHOW HORMONE SEASONALITY WITH ANTIPHASE AND SPRING SHIFT

To test such predictions requires accurate data on hormones collected in different months. Here we can utilize "big data." An exciting advance in systems medicine is the availability of large, searchable medical datasets, called **electronic health records (EHR)**. For example, Israel's largest health insurer (Clalit) allows researchers to study anonymized data on about half of the country's population over 15 years totaling about 50 million life-years, with broad socioeconomic and ethnic representation. The data includes disease codes, drugs purchased and blood tests. A playground to explore hypotheses and look for patterns.

Alon Bar and Avichai Tendler in their PhD used this data to explore hormone blood-test seasonality. They had to overcome a major challenge associated with medical datasets. This is **ascertainment bias** – medical tests are done for a clinical reason, as opposed to a uniform sample of the population. They addressed this together with Amos Tanay's group by filtering out data from people with medical conditions that can confound the results. For example, they filtered out, for each blood test, data from individuals that took a drug that affects that specific test. Identifying such drugs was done from the medical record data itself, by searching for drugs that significantly affect test results on average. If you want to know more and see how machine learning can be used to predict the risk for future diseases from one's medical record, see Cohen et al. (2021).

After controlling for such confounders, they plotted mean hormone levels as a function of the month of the year (Figure 3.16). Cortisol shows a peak in February and a minimum in August, and x_2 (ACTH) peaks in summer (Figure 3.16, orange line). The predictions hold – a spring shift and an antiphase.

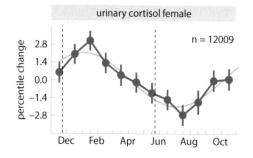
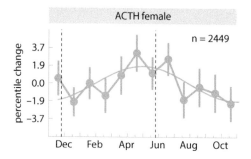

FIGURE 3.16 A large dataset of laboratory tests (Clalit) shows seasonal oscillations with a late winter peak of cortisol (left) and a summer peak of ACTH (right). The number of tests n is indicated. Adapted from Tendler et al. (2021).

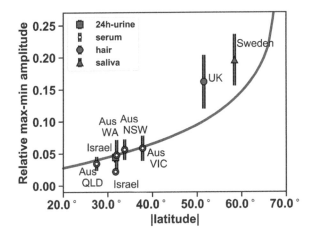

FIGURE 3.17 Cortisol seasonal amplitude rises with distance from the equator, as predicted by the model. Measurements in Australia (AUS) show a peak shifted by 6 months corresponding to Australian winter. Adapted from Tendler et al. (2021).

The oscillation amplitude is only a few percent. This matches the predicted amplitude at the 31° latitude of Israel. Seasonal input depends on geography because day length varies more the farther you are from the equator. The model hence predicts larger amplitudes at higher latitudes (Figure 3.17), reaching about 20% and 30% peak-to-trough in London and Stockholm, respectively. Tests of cortisol in other countries show the expected rise with latitude (Figure 3.17). Tests in Australia show the same seasonal dependence but shifted by 6 months, since Australian winter is in May–August.

Since the *HPA* axis is implicated in mood disorders, as discussed below, the seasonal variation in the *HPA* axis may contribute to "winter blues," officially called **seasonal component of affective disorder (SAD)**. This disorder is common at high latitudes, with 10% prevalence in Alaska versus 1% in Florida.

The two-gland feedback loop design of the HPA axis appears also in other hormonal systems controlled by the hypothalamus and pituitary, as we will see in the next chapter. These "HP" axes control reproduction, growth, and metabolism. Because all of these HP axes have a slow feedback loop between the glands, they should all show seasonal oscillations. Indeed, Clalit data shows seasonal oscillations in most hormones with amplitudes of 4%–10%. Some of these tests exceed 6 million data points(!), with error bars (s.e.m) smaller than the dots in Figures 3.18 and 3.19. The hormones in the different HP axes show the expected hallmarks, a spring shift in x_3 (such as thyroid hormone and the sex hormone estradiol) and an antiphase between x_3 and its upstream hormone x_2 (TSH and FSH), in which x_2 peaks in summer.

The peak of the effector hormones x_3 in late winter and early spring raises the possibility that human physiology has a peak season for reproduction, growth, and stress response. This is supported by data on late winter peaks in child growth, cognitive ability, and sperm quality. Cultural and regional effects complicate the interpretation of data, but overall, there is a case for a biological basis of human seasonality. Although modernity makes seasons less impactful, perhaps there is a season for every purpose.

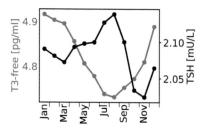

FIGURE 3.18 The Clalit dataset of blood tests shows antiphase between thyroid hormone T3 and its upstream hormone TSH. Adapted from Tendler et al. (2021).

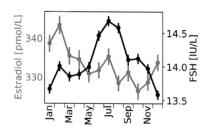

FIGURE 3.19 The Clalit dataset of blood tests shows antiphase between the sex hormone estradiol and its upstream hormone FSH. Adapted from Tendler et al. (2021).

BIPOLAR DISORDER HAS A TIMESCALE OF WEEKS-MONTHS, AND THIS TIMESCALE CAN BE GENERATED BY GLAND SIZE FLUCTUATIONS

Let's finish up by using our concepts to explore a mood disorder called **bipolar disorder** (Vieta et al. 2018). It is characterized by periods of months or years of depression with negative thoughts and low energy, and periods of mania for a week or more with excessive high energy, irritability, and delusions.

These mood episodes are not like familiar everyday mood swings. Depression seriously interferes with the ability to feel joy, function, eat, and get out of bed; mania can damage relationships and work. Bipolar disorder is a leading cause of suicide (Fardet et al. 2012) and afflicts about 1% of the worlds' population.

As in most psychiatric conditions, the biological understanding of bipolar disorder is still lacking. It has a strong genetic component, but no genes of large effect have been identified. Medications such as lithium can stabilize moods for some people, but the mechanism is not clear. Bipolar depression is hard to treat, and improving treatment is a huge unmet need.

There are also no blood tests or other objective biomarkers for bipolar disorder – or for any other psychiatric condition so far. This is unlike diabetes where a glucose test is the basis for diagnosis and follow-up of patients. Instead, diagnosis and follow-up of bipolar disorder is done by interviews with psychiatrists.

Since this disease is so poorly understood, we can afford to speculate using the concepts of this book. We will focus on the timescales of mood episodes in bipolar disorder. There is no doubt that bipolar disorder involves changes in brain circuits. Let's explore the hypothesis that the month-timescale is not due to brain changes in isolation from the rest of the body, but instead is due to the effects on the brain of hormone changes that take months due to the dynamics of gland size variations.

BIPOLAR DISORDER IS A STRESS-RELATED DISEASE

The glands of the *HPA* axis are natural candidates for this timescale. The *HPA* axis is dysregulated in major depression and bipolar disorder: about 50% of people with major depression have high cortisol and enlarged adrenal cortex, and people with bipolar disorder average about twice the mean cortisol levels of the undiagnosed population (Belvederi Murri et al. 2016).

High cortisol is not just a readout of psychiatric symptoms. High cortisol levels for weeks can also cause mood episodes (Belvederi Murri et al. 2016). Bipolar episodes are often preceded by stressful life events. High cortisol causes mania in about 30% of individuals with Cushing's syndrome in which a tumor secretes high levels of cortisol, and depression occurs in about 50% of these patients. High doses of steroids taken for a few weeks or longer, which as mentioned above are artificial versions of cortisol, cause mania symptoms in about 4% of patients, often followed by depression (Judd et al. 2014). The risk for mania rises with the dose of steroids (Figure 3.20). Thus a fraction of the population is susceptible to mood episodes caused by elevated cortisol.

We are ready to propose a mechanism for the timescales of bipolar disorder. We assume that people with bipolar disorder are within the fraction of the population susceptible to cortisol-induced mood episodes. But what can generate sufficiently high levels of cortisol for weeks? We assume that people with bipolar disorder have an additional neuropsychological trait called emotional hyper-reactivity in which they generate larger input signals to the *HPA* axis in response to life events. Such emotional hyper-reactivity is often reported for people with bipolar disorder (Dargél et al. 2020; Sperry and Kwapil 2022).

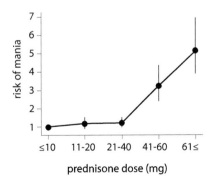

FIGURE 3.20 Risk of mania rises with steroid dose. Analysis of UK medical records shows that people who take cortisol analogue drugs such as prednisone for several weeks or longer show enhanced risk of mania after treatment begins compared to a matched control population. Normal cortisol level is equivalent to about 8mg prednisone. Adapted from Fardet et al. (2012).

The last piece in the puzzle is supplied by the HPA model. The HPA gland-mass feedback loop can amplify the daily stress inputs into large cortisol oscillations on the scale of months. These oscillations are larger in people with emotional hyper-reactivity because of their enhanced inputs to the *HPA* axis, increasing the chances of crossing the threshold to trigger mania and depression.

We can model the proposed mechanism by introducing to the HPA model a noisy input signal $u(t)$ that represents the daily stress inputs to the *HPA* axis. In people with bipolar disorder, let's assume that this noisy input is larger than in people without bipolar disorder because of emotional hyperreactivity.

Such a noisy input to the *HPA* two-gland feedback loop causes a fascinating phenomenon: Noisy oscillations of gland sizes and hormones, with a typical timescale of many months. The larger the noise, the larger the amplitude of these oscillations. To see this, put the spring in its honey container on the back of a flatbed truck driving on a rough gravel road (Figure 3.21). The spring picks up vibrational frequencies close to its natural (resonance) frequency and starts making noisy, erratic oscillations. This is exactly what happens to the HPA glands and hormones (Figure 3.22). The rougher the road (more noise), the larger the oscillations.

These noisy oscillations look like a messy ring around the fixed point in the phase portrait (Figure 3.23, red trajectory). In fact, any system with a spiral fixed point – arrows that

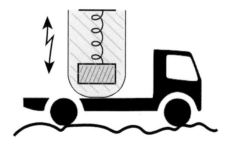

FIGURE 3.21 The HPA axis with noisy input is analogous to a mass on a spring in a viscous fluid container placed on the back of a truck driving on a gravel road. The spring shows noisy oscillations at its natural frequency.

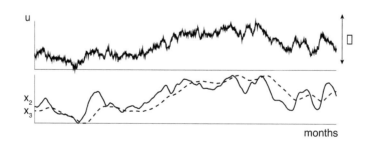

FIGURE 3.22 The HPA hormones in the model show noisy fluctuations on the scale of months in response to a white noise input.

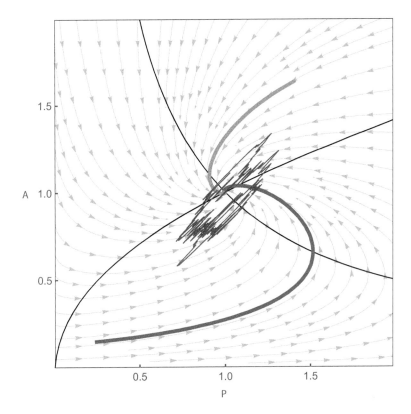

FIGURE 3.23   Phase portrait of the HPA axis glands with a white-noise input shows a noisy oscillation around the spiral fixed point.

spiral into the fixed point instead of going directly into it – will exhibit noise-driven oscillations[3] (Alon 2019).

Note that these oscillations are different from the seasonal oscillations discussed above. Seasonal oscillations are tiny. To detect them, one needs to average over thousands of people. The noisy oscillations we are talking about now are much larger and are not synchronized with the seasons.

The HPA model with a white-noise input produces cortisol fluctuations with a timescale of many months to a year, despite the fact that the noise input is merely random stresses on the timescale of days (Figure 3.24). One way to visualize these year scale fluctuations is using a Fourier transform, which decomposes a signal to a sum of oscillations with different frequencies. The higher the Fourier amplitude, the more dominant that frequency component in the signal. The simulated cortisol shows dominant low frequencies corresponding to a year timescale (Figure 3.24). Simulated bipolar patients, namely simulations with larger white noise input, show even higher amplitudes at the year scale.

To test the predicted cortisol fluctuations, we did an experiment that used hair to measure cortisol over a year. Cortisol diffuses from the circulation into the hair follicle and binds the hair inner cortex, where it sits passively. This is a phenomenon with no physiological consequence that provides a cool way to measure the hormone over time. Since

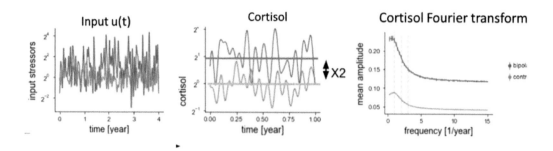

FIGURE 3.24   Simulations of the HPA model with white-noise input. The input $u(t)$ varies rapidly and randomly (left). Bipolar disorder is modelled using larger white noise. Cortisol fluctuates with time (center).Fourier analysis shows that low frequency oscillations are dominant, with periods of about a year, and have higher amplitude in simulated bipolar disorder than in control simualtions (right).

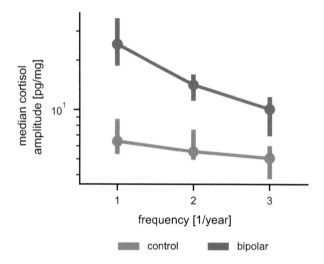

FIGURE 3.25   Fourier amplitudes of hair cortisol measured over 12 months. People with bipolar disorder ($n=26$) showed higher mean cortisol level and higher amplitude of year-scale fluctuations than control participants ($n=59$). Adapted from Milo et al. (submitted).

each centimeter of hair is approximately a month of growth, a 12 cm hair sample contains a record of the cortisol history over a year.

Both control participants and people with bipolar disorder showed fluctuations with a dominant frequency of a year as predicted by the model (Maimon et al. 2020); people with bipolar disorder had higher amplitude fluctuations that can easily reach 5-fold higher levels than the normal baseline (Figure 3.25).

Such prolonged periods of high cortisol is just what sets off mania and depression in people taking large doses of glucocorticoids for many weeks.

These are data from a small sample of people and need to be verified by additional studies. Still, we can hypothesize that the noisy oscillations of the *HPA* axis might account for some of the inertia and timescale of bipolar mood episodes (Figure 3.25)[4].

If such HPA oscillations turn out to be causal for bipolar disorder, and not merely a read-out, they suggest pathways for treatment. A drug that can reduce cortisol should reduce the risk of mania and depression in people susceptible to bipolar disorder. Such a drug can be helpful more generally for people in chronic stress, because high cortisol increases the risk of heart disease and diabetes.

THE HPA CIRCUIT SUGGESTS DRUG TARGETS TO LOWER CORTISOL

Which drug is likeliest to lower cortisol? We can use the HPA circuit model to screen for drug targets. A drug is a molecule that affects a certain parameter in the system – inhibiting a receptor for example. In a "circuit-to-target" approach, we systematically change each model parameter and look for those that can lower steady-state cortisol.

This is important, because certain drug candidates for the *HPA* axis held great hope in the 1990s and 2000s when large clinical trials were performed. Hope was followed by disappointment – the drugs had little long-term effect on depression or bipolar disorder in clinical trials. This includes a drug that blocks the cortisol receptor, hoping to reduce the effect of high cortisol.

The HPA model can explain the failure of most drugs. Due to the compensation properties of the circuit, most parameter changes do not, in fact, lower the cortisol steady state. In other words, **most of the possible drugs would not work because of compensation!** This is due to the robustness of the cortisol steady state – its dependence on only a very few parameters – as seen by recalling the steady-state formula

(3.12)
$$x_3^{st} = \frac{q_1 b_P}{\alpha_1 a_P} u$$

Therefore, most of the possible HPA-targeting drugs, like receptor inhibitors, hormone production inhibitors or hormone analogous, simply would not change cortisol in the long term. The glands change their size to compensate fully.

For example, blocking the cortisol receptor releases the inhibition of cortisol on the production of x_1 and x_2, making the adrenal gland grow and release more cortisol, exactly compensating for the blocking of its receptors. The drug thus has an effect for a few weeks, but stops working when the circuit compensates.

The HPA model can help us pinpoint the few intervention points that work to lower cortisol. One of these points is reducing the input u. This can be done by psychological and well-being interventions that reduce stress. There are two other intervention points suitable for drugs. Cortisol x_3^{st} can be reduced according to Eq. 3.12 by lowering the parameter b_P, namely the effect of x_1 (CRH) on P cell growth. This can be done by inhibiting the CRH-receptor in P cells using a receptor antagonist. Another way to lower cortisol is to increase the CRH removal-rate α_1, for example by antibodies that bind and block CRH (Futch et al. 2019). Such drugs may be candidates for lowering cortisol and treating mood disorders.

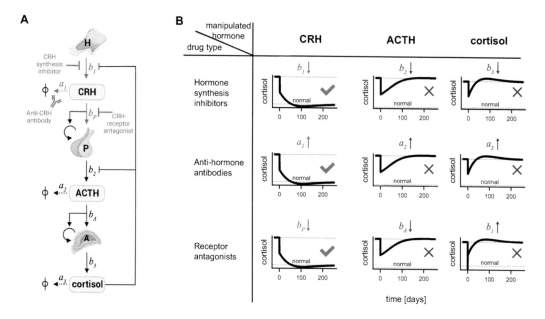

FIGURE 3.26 HPA model suggests targets for drugs that can lower long-term cortisol. (A) Cortisol dynamics are shown following a change in one of the model parameters. (B) These changes simulate drugs that block synthesis, inhibit receptors, or enhance removal of each of the three HPA hormones. Only drugs with a green check mark reduce the long-term cortisol steady state. For the cortisol receptor inhibitor, cortisol effect is plotted, namely cortisol bound to the receptor.

The simulated effects of all possible synthesis inhibitors, receptor antagonists, and anti-hormone antibodies for the three hormones in the *HPA* axis are shown in Figure 3.26. Only the drugs with green check marks can lower steady-state cortisol in the long term. Without a circuit model, we run the risk of developing the wrong drugs.

SUMMARY

To sum up, when two organs control each other's sizes, oscillatory phenomena on the timescale of months can occur. The two-gland oscillator allows the HPA stress pathway to synchronize with the seasons providing hormonal set-points for different times of the year. It explains the months-long aftermath of recovery from stress, illness, or addiction. A cost of this oscillator may be noisy mood swings, setting off mania and depression in a small percentage of people. The gland-mass changes make it hard to drug the pathways – most potential drugs are compensated away and do not work – and the circuits can help to predict which drugs might work. More generally, the body can be considered as an ensemble of interacting organs that constantly adjust their sizes and activities to adapt to changing conditions and to the states of the other organs.

To reduce stress, let's take a nice deep sigh of relief.

EXERCISES

Exercise 3.1: The HPA circuit

The goal of this exercise is to flex your algebra muscles, and to see the mathematical origin of robustness. The *HPA* axis (Figure 3.27) can be described by the following equations:

(3.13)
$$\frac{dx_1}{dt} = q_1 \frac{u}{x_3} - a_1 x_1$$

(3.14)
$$\frac{dx_2}{dt} = q_2 \frac{x_1}{x_3} P - a_2 x_2$$

(3.15)
$$\frac{dx_3}{dt} = q_3 x_2 A - a_3 x_3$$

(3.16)
$$\frac{dP}{dt} = P(b_P x_1 - a_P)$$

(3.17)
$$\frac{dA}{dt} = A(b_A x_2 - a_A)$$

a. Explain the meaning of each parameter. For example, b_p is the *P* cell proliferation rate per unit CRH, x_1.

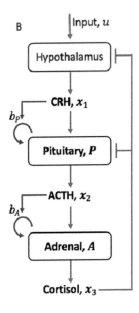

FIGURE 3.27   HPA circuit with gland-mass changes.

b. Fast timescale steady state: Solve the steady state of Eqs. 3.13–3.15, assuming that the gland masses A and P are constant. What are the quasi-steady-state values of the hormone concentrations x_1, x_2, and x_3?

c. Full solution steady state: Solve the steady state of the full model, Eqs. 3.13–3.17. Start with Eqs. 3.16 and 3.17, to show that x_1 and x_2 must reach specific values. Then use these values in Eqs. 3.13–3.15 to find the steady states of x_3, A, and P.

d. Compare the solutions of b and c. Which solution has hormone levels that are more robust (insensitive) to variations in model parameters? Explain biologically what mechanism provides this robustness (50 words).

Exercise 3.2: Steroid withdrawal

Many medical situations are treated by drugs that mimic cortisol called corticosteroids (such as dexamethasone or prednisone), to reduce inflammation, swelling, and autoimmune diseases. Millions of people take such drugs at high doses for many weeks.

a. Model such a drug by adding its dose D to x_3 in the inhibitory terms in the Eqs. 3.13 and 3.14, so that the $1/x_3$ terms become $1/(x_3 + D)$. Explain.

b. Solve Eqs. 3.13–3.17 for the effect of the drug on the steady-state hormone levels. What is the effect on the gland sizes?

   In (a) you should see that compensation is possible only until a certain value of D. What is this value of D? What happens when D exceeds this value? Hint: Hormone levels and gland masses cannot be negative.

c. Why is it dangerous to stop taking the drug all at once? This effect is called steroid addiction or steroid withdrawal. (50 words).

Exercise 3.3: Numerical simulation of chronic stress

a. Let's model the impact on the HPA axis of high-stress input for a year. Numerically simulate the HPA Eqs. 3.1–3.5 for a step change in which $u = 1$ goes to $u = 2$ at time $t = 0$. Run the simulation until the model reaches its new steady state. Use $q_1 = q_2 = q_3 = b_P = b_A = 1$, $a_1 = 1/5\,\text{min}$, $a_2 = 1/30\,\text{min}$, $a_3 = 1/90\,\text{min}$ and $a_P = a_A = 1/60\,\text{days}$. As initial conditions use the steady-state values of the system with $u = 1$ from Solved Exercise 3.1.

b. What happens to the levels of the three hormones after this step? Does x_3 behave differently from the other two hormones? Does this make sense biologically?

c. Suppose that the crisis is suddenly over. Simulate the effect of a down-step of u from $u = 2$ to $u = 1$ at time $t = 1$ year. What do you observe happens to the hormones and the glands? How long does it take the gland masses to return to their original state?

What might you conclude about how long it takes to fully recover from a long stress period? (50 words)

d. Brief crisis: supposed there is a brief crisis that lasts 1 week. Start at steady state at $u=1$ and simulate a pulse of u that rises to $u=2$ and then back to $u=1$. Compare the impact of the brief crisis on the glands and hormones to the prolonged stress of (a–c), both during the stress pulse and in the recovery period. What explains the difference?

Exercise 3.4: Circuit to target

In this exercise, we use the HPA equations of Exercise 3.1 to screen for drug targets to lower cortisol – on the back of an envelope. In each of the following, solve analytically and simulate the response to a 2-fold reduction in the relevant parameters over 6 months.

a. Synthesis inhibitor drugs lower the production parameters q_1, q_2, or q_3 when they inhibit the synthesis of x_1, x_2, or x_3, respectively. What is the effect of the three possible synthesis inhibitors on the cortisol steady-state level? Explain in terms of compensation by gland-mass changes.

b. Receptor antagonists block the effects of a hormone on its target glands. For example, a CRH receptor antagonist reduces the effects of x_1 (CRH), lowering the parameters q_2 and b_p. What is the effect of the three possible receptor antagonists on the cortisol steady-state level? Explain in terms of compensation by gland-mass changes.

c. Hormone inhibitors such as anti-hormone antibodies bind the hormones and effectively increase their removal rates. What is the effect of the three possible inhibitors on the cortisol steady state? Explain in terms of compensation by gland-mass changes.

Exercise 3.5: Nullclines and phase portraits

To enhance our skill with nullclines (lines where one variable has zero derivative, namely $dA/dt=0$ or $dP/dt=0$ in the HPA model), note how in Figure 3.12 the arrows that cross each nullcline are always either horizontal or vertical.

a. Explain why in Figure 3.12 on the rising nullcline the arrows are horizontal, while on the declining nullcline the arrows are vertical.

b. Analyze in a similar way the phase portrait for the insulin-glucose circuit in Chapter 1

c. Explain why each nullcline separates the phase portrait into two regions, such that in one region a variable always rises and in the other it always falls.

d. Draw two fanciful nullcline curves for two variables x and y, such that the nullclines that cross at three points. It is given that two of these fixed points are stable and one is unstable. Sketch the arrows for this system, using the properties of nullclines discussed in this exercise.

NOTES

1 The middle of the adrenal gland secretes (and gives its name to) adrenaline. It is not part of the HPA axis. The adrenal cortex has three layers that make three hormones – cortisol, androgen and aldosterone. The variable A in our circuit refers to the mass of the layer that secretes cortisol.

2 More generally, binding of cortisol inhibits production as $1/(K + x3)$ where K is the binding affinity to the receptor, an equation called the Michalis-Menten equation. The $1/x3$ term is a good approximation when $x3$ concentration exceeds K, as it does for cortisol. Blood levels of cortisol are on the order of 100 nM, while K is on the order of 10 nM for the low-affinity mineralocorticoid receptor. Another cortisol receptor with a higher K, the glucocorticoid receptor, comes into play at very high cortisol levels, and plays a role in depression.

3 For experts in linear analysis, the frequency of the noise-driven oscillations equals the frequency of the spiral, determined by the imaginary part of the eigenvalues at the fixed point.

4 Stepping back for a moment from bipolar disorder, we may ask why we have moods in the first place? One hypothesis is that moods evolved as a mechanism to allocate effort in proportion to reward (Stearns and Medzhitov 2016). Quoting Nesse (Fried and Nesse 2015; Buss 2005): "When payoffs are high, a positive mood increases initiative and risk taking. When risks are substantial or effort is likely to be wasted, low mood blocks investment." The inertia of HPA gland sizes can help to integrate risks and rewards over months, without being fooled by an outlier positive stimulus into thinking that things are going well. This design has a fragility to mood disorders including bipolar disorder.

REFERENCES

Alon, Uri. 2019. "An Introduction to Systems Biology." *An Introduction to Systems Biology*, July. https://doi.org/10.1201/9780429283321.

Belvederi Murri, M., D. Prestia, V. Mondelli, C. Pariante, S. Patti, B. Olivieri, C. Arzani, M. Masotti, M. Respino, M. Antonioli, L. Vassallo, G. Serafini, G. Perna, M. Pompili, M. Amore. 2016. "The HPA Axis in Bipolar Disorder: Systematic Review and Meta-analysis." *Psychoneuroendocrinology* 63: 327–42. doi: 10.1016/j.psyneuen.2015.10.014.

Buss, David M. 2005. *The Handbook of Evolutionary Psychology*. Hoboken, NJ: John Wiley & Sons, Inc.

Byyny, Richard L. 2009. "Withdrawal from Glucocorticoid Therapy." Review-article. Massachusetts Medical Society. World. November 16, 2009. https://doi.org/10.1056/NEJM197607012950107.

Cohen, Netta Mendelson, Omer Schwartzman, Ram Jaschek, Aviezer Lifshitz, Michael Hoichman, Ran Balicer, Liran I. Shlush, Gabi Barbash, and Amos Tanay. 2021. "Personalized Lab Test Models to Quantify Disease Potentials in Healthy Individuals." *Nature Medicine* 27 (9): 1582–91. https://doi.org/10.1038/s41591-021-01468-6.

Dargél, Aroldo A., Stevenn Volant, Elisa Brietzke, Bruno Etain, Emilie Olié, Jean-Michel Azorin, Sebastian Gard, et al. 2020. "Allostatic Load, Emotional Hyper-Reactivity, and Functioning in Individuals with Bipolar Disorder." *Bipolar Disorders* 22 (7): 711–21. https://doi.org/10.1111/bdi.12927.

Dixon, Rosina B., and Nicholas P. Christy. 1980. "On the Various Forms of Corticosteroid Withdrawal Syndrome." *The American Journal of Medicine* 68 (2): 224–30. https://doi.org/10.1016/0002-9343(80)90358-7.

Fardet, Laurence, Irene Petersen, and Irwin Nazareth. 2012. "Suicidal Behavior and Severe Neuropsychiatric Disorders Following Glucocorticoid Therapy in Primary Care." *American Journal of Psychiatry* 169 (5): 491–97. https://doi.org/10.1176/appi.ajp.2011.11071009.

Fried, Eiko I., and Randolph M. Nesse. 2015. "Depression Sum-Scores Don't Add up: Why Analyzing Specific Depression Symptoms Is Essential." *BMC Medicine* 13 (1): 1–11. https://doi.org/10.1186/S12916-015-0325-4/TABLES/1.

Futch, H.S., K.N. McFarland, B.D. Moore, M.Z. Kuhn, B.I. Giasson, T.B. Ladd, K.A. Scott, M.R. Shapiro, R.L. Nosacka, M.S. Goodwin, and Y. Ran. 2019. "An Anti-CRF Antibody Suppresses the HPA Axis and Reverses Stress-Induced Phenotypes." *Journal of Experimental Medicine* 216(11): 2479–2491.

Graber, A. L., R. L. Ney, W. E. Nicholson, D. P. Island, and G. W. Liddle. 1965. "Natural History of Pituitary-Adrenal Recovery Following Long-Term Suppression with Corticosteroids." *The Journal of Clinical Endocrinology & Metabolism* 25 (1): 11–16. https://doi.org/10.1210/JCEM-25-1-11.

Judd, Lewis L., Pamela J. Schettler, E. Sherwood Brown, Owen M. Wolkowitz, Esther M. Sternberg, Bruce G. Bender, Karen Bulloch, et al. 2014. "Adverse Consequences of Glucocorticoid Medication: Psychological, Cognitive, and Behavioral Effects." *American Journal of Psychiatry* 171 (10): 1045–51. https://doi.org/10.1176/appi.ajp.2014.13091264.

Karin, Omer, Moriya Raz, Avichai Tendler, Alon Bar, Yael Korem Kohanim, Tomer Milo, and Uri Alon. 2020. "A New Model for the HPA Axis Explains Dysregulation of Stress Hormones on the Timescale of Weeks." *Molecular Systems Biology* 16 (7): e9510. https://doi.org/10.15252/msb.20209510.

Karin, Omer, Moriya Raz, and Uri Alon. 2021. "An Opponent Process for Alcohol Addiction Based on Changes in Endocrine Gland Mass." *IScience* 24 (3): 102127. https://doi.org/10.1016/j.isci.2021.102127.

Maimon, Lior, Tomer Milo, Rina S. Moyal, Avi Mayo, Tamar Danon, Anat Bren, and Uri Alon. 2020. "Timescales of Human Hair Cortisol Dynamics." *IScience* 23 (9): 101501. https://doi.org/10.1016/j.isci.2020.101501.

Nicolaides, Nicolas C., Zoi Galata, Tomoshige Kino, George P. Chrousos, and Evangelia Charmandari. 2010. "The Human Glucocorticoid Receptor: Molecular Basis of Biologic Function." *Steroids* 75 (1): 1–12. https://doi.org/10.1016/j.steroids.2009.09.002.

Sperry, Sarah H. and Thomas R. Kwapil. 2022. "Bipolar Spectrum Psychopathology Is Associated with Altered Emotion Dynamics Across Multiple Timescales." *Emotion* 22(4): 627.

Stearns, Stephen C, and Ruslan Medzhitov. 2016. "Stephen C. Stearns, and Ruslan Medzhitov. 2016. "Evolutionary Medicine." *Evolution, Medicine, and Public Health* 2016 (1): 69–70. https://doi.org/10.1093/EMPH/EOW008.

Tendler, Avichai, Alon Bar, Netta Mendelsohn-Cohen, Omer Karin, Yael Korem Kohanim, Lior Maimon, Tomer Milo, et al. 2021. "Hormone Seasonality in Medical Records Suggests Circannual Endocrine Circuits." *Proceedings of the National Academy of Sciences* 118 (7): e2003926118. https://doi.org/10.1073/pnas.2003926118.

Vieta, Eduard, Michael Berk, Thomas G. Schulze, André F. Carvalho, Trisha Suppes, Joseph R. Calabrese, Keming Gao, Kamilla W. Miskowiak, and Iria Grande. 2018. "Bipolar Disorders." *Nature Reviews Disease Primers* 4 (1): 1–16. https://doi.org/10.1038/nrdp.2018.8.

Vinther, Frank, Morten Andersen, and Johnny T. Ottesen. 2011. "The Minimal Model of the Hypothalamic-Pituitary-Adrenal Axis." *Journal of Mathematical Biology* 63 (4): 663–90. https://doi.org/10.1007/S00285-010-0384-2.

The Thyroid and Its Discontents

THE HPA AXIS WE just studied is one of four hypothalamic-pituitary axes, or HP axes, that control pillars of life – growth, reproduction, stress, and metabolism (Figure 3.5.1). Each axis has its own circuit similar to the HPA axis (Figure 3.5.2).

We now focus on the *HP*-thyroid axis, or **HPT axis** for short, that controls metabolic rate. The thyroid axis is biologically fascinating and is the target of several common diseases – especially diseases in which the body attacks itself, called autoimmune diseases.

This brief chapter is the first of a two-chapter mini-series, where we understand the origin of autoimmune diseases: what is the logic of the body attacking itself. We use the thyroid to demonstrate the biology and treatment of such diseases. We also revisit some of the principles that we studied to understand dynamical features of these diseases.

THE *HPT* AXIS IS DESIGNED TO KEEP A STEADY LEVEL OF THYROID HORMONE T4

The thyroid is a 20-g gland at the front of our throat shaped like a butterfly (Figure 3.5.3). It secretes thyroid hormone, T4, that affects all cells and has far-reaching effects on the heart and on metabolism. It is the effector hormone of the *HPT* axis (Figure 3.5.4). This axis, like the *HPA* axis, is a cascade of three hormones. The hypothalamus *H* secretes hormone x_1 (thyroid releasing hormone – TRH) that makes the pituitary *P* secrete hormone x_2 (thyroid stimulating hormone – TSH) from pituitary thyrotroph cells. TSH causes the thyroid to secrete T4, our x_3 hormone. As in the *HPA* axis, the effector hormone x_3 inhibits the production of its two upstream hormones x_2 and x_1, TSH and TRH (Figure 3.5.4).

There is a further similarity with the *HPA* axis – the hormones also control the sizes of the glands. TSH makes the thyroid cells proliferate and grow. In fact, excessive growth of the thyroid is infamous, called goiter. It occurs when iodine, essential for making T4, is low in the diet, as in areas far from the sea which in old times lacked access to the iodine in

 DOI: 10.1201/9781003356929-7

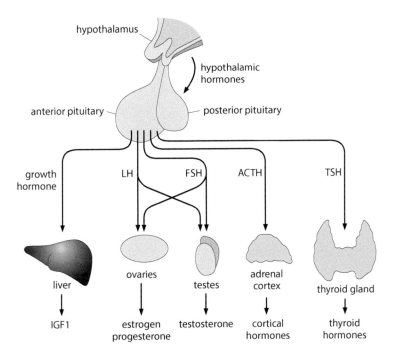

FIGURE 3.5.1 The *HP* axes regulate growth, reproduction, stress response, and metabolism. Each axis has its own x_1 hormones from the hypothalamus that induce a second x_2 hormone from the pituitary which activates secretion of the effector hormone x_3 from an effector gland.

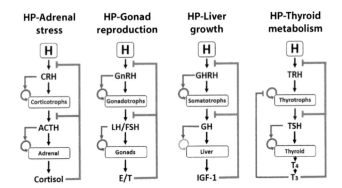

FIGURE 3.5.2 Each *HP* axis has a circuit in which the effector hormone inhibits the production of its upstream hormones (gray arrow). The hormones also act as growth factors for their downstream glands (red arrows). The *HP*-thyroid axis has a slightly different circuit. The names of the hormone-secreting pituitary cell types are indicated.

seafood. Low T4 releases the inhibition of TSH. High TSH makes the thyroid grow by up to a factor of 10, like a small melon at the throat, to try to make enough thyroid hormone.

Due to the inhibition of TSH by T4, the two hormones show an inverse relationship – when T4 is low TSH is high (Figure 3.5.5). The relation between TSH and T4 is steep, almost exponential. TSH is a sensitive indicator of problems with the thyroid. It varies by a factor of a thousand whereas T4 varies far less, with a normal range of 10–20 pmol/L. That is why

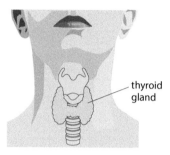

FIGURE 3.5.3 The thyroid is a 20 g gland at the front of the throat.

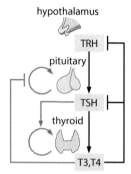

FIGURE 3.5.4 In the thyroid axis, TSH is the growth factor for the thyroid. The effector hormones T4 and T3 inhibit the growth of pituitary cells that secrete TSH.

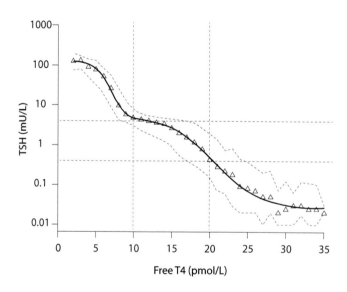

FIGURE 3.5.5 TSH and thyroid hormone T4 show an inverse relationship that is approximately log-linear. The normal ranges for the hormones are indicated by dashed lines. Adapted from Hadlow et al. (2013).

TSH is such a common blood test. It can even tell if there will be future problems, even though now T4 is fine. Such abnormal TSH and normal T4 is called a **subclinical disease**, because it has no symptoms, but is a warning sign.

STRUCTURE-FUNCTION RELATION IN THE HPT CIRCUIT

The pituitary cells that secrete TSH also have size control – the effector hormone T4 inhibits P cell growth. This is a variation on the other axes like the *HPA* axis – in the other axes, x_1 enhances P growth instead of x_3 inhibiting it. Thus, the HPA and HPT circuits differ by a single interaction arrow – the control of P growth (Figure 3.5.2).

Why this different circuit design? The answer is that the circuit structure of each axis fulfills the specific function required by that axis. The HPT axis performs homeostasis by keeping nearly constant T4 levels, whereas the *HPA* axis is an input-response circuit, designed to produce a wide range of x_3 (cortisol) concentrations according to stress inputs. Indeed, free T4 levels in the circulation are constant to within a factor of 2. We say "free T4" because most of T4 is bound in the blood to carrier proteins and is not active. As mentioned above, the free T4 range in the population is 10–20 pmol/L. Each healthy individual varies over time by about 50% of this range around their own personal set point.

To see how the HPT circuit achieves homeostasis, consider the equation for pituitary mass P in the *HPT* axis

$$(3.5.1) \qquad dP/dt = P\left(b_p/x_3 - a_p\right)$$

note the b_p/x_3 term which arises because x_3 inhibits P growth. The only way to achieve steady state $dP/dt = 0$ is for the terms in the parentheses to equal zero. This locks thyroid hormone x_3 to a steady state equal to the ratio of P cell production and removal parameters

$$(3.5.2) \qquad x_{3,st} = b_p/a_p$$

This thyroid hormone steady state is robust to changes in all other physiological parameters – reminiscent of the homeostasis of glucose levels in Chapter 2.

In contrast, in the HPA axis, pituitary growth is controlled by x_1 so that $dP/dt = P(b_P\, x_1 - a_P)$. This equation locks x_1 to a constant steady-state $x_1 = a_P/b_P$ and keeps $x_{3,st}$ free to vary with input u, as we saw in Chapter 3, as appropriate for a stress response pathway which is sensitive to the magnitude of the stressors.

Thus, circuit structure matches function.

HOW IS THYROID HORMONE PRODUCED?

For our understanding of diseases, we need some details of how thyroid hormone T4 is made in the thyroid. T4 is a molecule made of two carbon hexagonal rings taken from the amino acid tyrosine (Figure 3.5.6). Each ring has two iodines attached, totaling four iodines and hence the name T4. In the cells of the body, T4 is converted to a more active form T3, so called because one of four iodines is removed by iodinase enzymes.

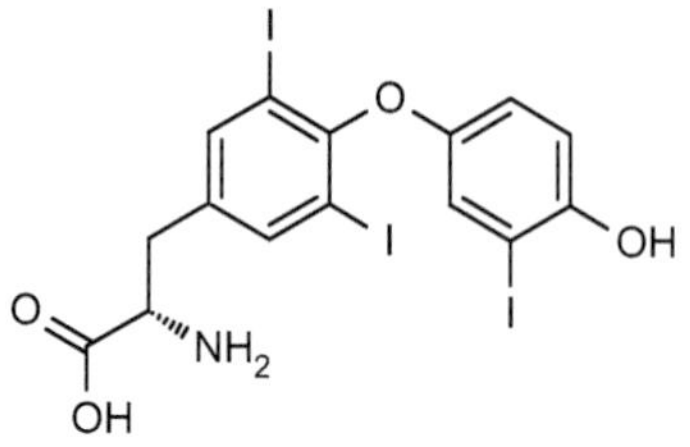

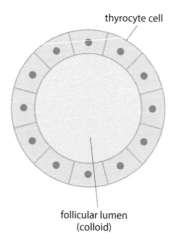

FIGURE 3.5.6   Thyroid hormone T4 is made of a carbon backbone bound to four iodine atoms.

thyrocyte cell

follicular lumen
(colloid)

FIGURE 3.5.7   Thyroid hormones are produced in thyroid follicles.

T4 is made in modular chemical factories in the thyroid called follicles (Figure 3.5.7). Each follicle is a layer of cells called thyrocytes surrounding a spherical pore filled with colloid, with a total size of about 400 μm. The thyrocytes import iodine from the blood vessels surrounding the follicle and then export the iodine into the colloid (Figure 3.5.8). They make huge amounts of a protein called *Tg* that has many tyrosines and dump this protein also into the colloid. There, iodine is added to the tyrosines on *Tg* by an enzyme called thyroid peroxidase, or TPO. Then, the cells import the iodinated *Tg* back inside, break it up into small pieces, and extract the T4, and export it out of the cell into the circulation.

Remember the key proteins *Tg* and TPO for future use. The upstream hormone TSH increases all of the above steps, as well as the production of *Tg*, TPO, and other transporters and enzymes involved in hormone production and secretion.

HYPERTHYROIDISM AND HYPOTHYROIDISM

To understand the function of thyroid hormone T4, let's see what happens when there is too much or too little of it. Diseases with too much T4 result in **hyperthyroidism**. The heart beats fast and irregularly, which can lead to dangerous arrhythmias. Hyperthyroidism increases metabolic rate – you eat more but lose weight. Emotions and thoughts race, sometimes resembling mania. There is a feeling of heat and sweat, due to thermogenesis. Muscles hurt. The hands might shake due to effects on the nervous system.

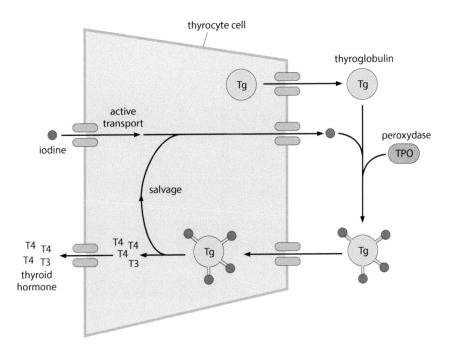

FIGURE 3.5.8 Thyroid hormones are produced by thyrocyte cells. The hormone precursor, thyroglobulin (*Tg*), is a protein with many copies of the carbon backbone of thyroid hormone and is secreted into the lumen of the follicle. Iodine is attached to *Tg* in the lumen by thyroid peroxidase (TPO). Iodinated *Tg* is transported back to the cell and degraded to make T4 and T3 which are transported into the circulation.

Opposite effects occur when thyroid hormone is too low, called **hypothyroidism**. The heart beats slowly, and you can be out of breath when climbing the stairs. You gain weight despite not eating more, with reduced appetite and constipation. Emotions tend toward depression. There is a clammy feeling of cold. In infants, hypothyroidism can lead to impaired development of the brain – that is why a TSH test at birth is so important.

HASHIMOTO'S DISEASE, A CASE OF SELF-ATTACK

The most common cause of hypothyroidism is Hashimoto's disease (Figure 3.5.9) (Chaker et al. 2022). It affects about 2% of the population, primarily women. In Hashimoto's disease, white blood cells called *T*-cells attack and kill thyrocytes.

The *T* cells normally attack virus infected cells, not healthy cells of the body. Unfortunately, in Hashimoto's, *T* cells attack healthy thyrocytes and damage the thyroid. More in the next chapter.

The antigens recognized by *T* cells in Hashimoto's are pieces of the *Tg* and TPO proteins discussed above. The *T* cells also activate *B* cells to make antibodies against *Tg* and TPO. The antibodies participate in damaging the thyroid. Hashimoto's is clinically identified by anti-*Tg* and anti-TPO antibodies in the blood.

Sometimes, early stages of Hashimoto's cause T4 to spill from destroyed cells, causing hyperthyroidism. This is soon followed by hypothyroidism as the thyroid is destroyed.

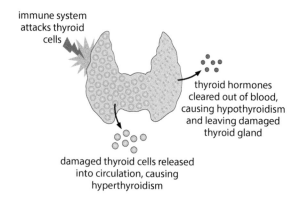

FIGURE 3.5.9 Hashimoto's thyroiditis, where *T* cells destroy thyrocytes and eventually destroy so much of the thyroid that thyroid hormone levels drop below their normal range – a serious condition known as hypothyroidism.

Hashimoto's disease is treated by supplying thyroid hormone in pills. The pills are taken for life since the autoimmune attack never allows the thyroid to recover. Most people live fine with these pills. However, a small percentage of the population has problems adjusting the dose. This causes problems for millions of people.

TSH SHOWS A DELAY AFTER HASHIMOTO'S DISEASE IS TREATED

When Hashimoto's is treated by T4 pills at the proper dose, T4 recovers to its normal range. TSH however remains high for 6 weeks after T4 normalizes (Figure 3.5.10). This delay is usually not explained in endocrinology textbooks. It cannot be explained by the hormone lifetimes which are much faster. This delay makes it difficult to adjust treatment dose and stabilize patients based on TSH blood tests.

The origin of the delay cannot be understood unless we take the changes of gland masses into account. When thyroid hormone production is compromised by autoimmune attack of the thyroid, there is less T4. Since T4 inhibits pituitary growth, low T4 lifts this inhibition. The pituitary is thus enlarged, to make more TSH to stimulate the thyroid and compensate for the killing. When the disease is treated, by taking thyroid hormone pills,

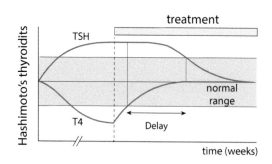

FIGURE 3.5.10 Delay in the treatment of Hashimoto's thyroiditis. When treated by pills that supply thyroid hormones, the levels of thyroid hormone T4 return to normal. TSH returns to normal only after a delay of about 6 weeks relative to T4.

it takes the enlarged pituitary 6 weeks to shrink back to its normal size. As a result, it takes TSH 6 weeks to return to baseline (Korem Kohanim et al. 2022).

SUBCLINICAL HASHIMOTO'S IS DUE TO DYNAMICAL COMPENSATION

In fact, the disease can go quietly for decades with no symptoms, long before T4 levels drop below normal (Figure 3.5.11). In this **subclinical** stage, thyroid cells are killed by the immune system. TSH is high but T4 is normal. Doctors recommend a follow-up test to see if T4 levels fall below their normal range.

The origin of the subclinical stage is the dynamical compensation effect that we discussed before. The pituitary cell total mass adjusts itself so that more TSH is secreted, to precisely compensate for the killing of thyrocytes.

To understand the subclinical stage in terms of equations, we can build on our experience with the *HPA* axis. The growth of thyroid mass T is stimulated by TSH, denoted $x2$, so that $dT/dt = T(b_T \, x2 - a_T)$. The steady state of this equation locks TSH to a steady state concentration, $x2 = a_T/b_T$, proportional to the removal rate of thyroid cells, a_T. Since autoimmune killing increases thyrocyte removal rate a_T, it causes the high TSH levels observed in Hashimoto's thyroiditis. The pituitary mass grows to supply this extra TSH. However, the level of thyroid hormone, denoted x_3, is kept constant by the previously mentioned equation for the pituitary mass, $dP/dt = P(b_P/x3 - a_P)$, so that thyroid hormone is normal $T4 = x3 = a_P/b_P$.

But according to law 2, biological processes saturate. If the killing rate of the autoimmune disease exceeds a threshold, the pituitary mass approaches its carrying capacity. Compensation by the pituitary maxes out. Any rise in autoimmune killing rate now causes the thyroid hormone x_3 to drop below its healthy range. Clinical hypothyroidism emerges. Many people with subclinical disease therefore rely on compensation by gland-mass changes for health.

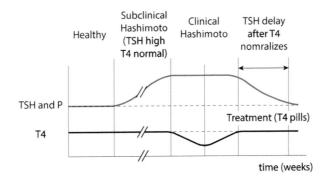

FIGURE 3.5.11 Subclinical Hashimoto's thyroiditis occurs when TSH is higher than normal but thyroid hormones are normal. Thyroid hormone is maintained in the normal range due to dynamic compensation in which the pituitary cells that secrete TSH grow in effective mass, producing excess TSH to compensate for autoimmune killing of the thyroid. When the pituitary cell mass approaches its carrying capacity, this compensation maxes out, and thyroid hormones drop below normal, instigating clinical Hashimoto's thyroiditis.

HYPERTHYROIDISM DUE TO TOXIC NODULES: A CASE OF A HYPER-SENSING MUTANT

We now turn to hyperthyroidism, too much T4. One major cause is a batch of growing cells in the thyroid that form a nodule that secretes too much thyroid hormone T4 (Figure 3.5.12). These nodules can be imaged by ultrasound, felt by touch, and seen to be active in secreting T4 by a radioactive iodine scan. Since they show up on the scan, they are called "hot nodules" or **toxic nodules**. They occur in about 1% of the population, primarily at old age.

There are also "cold nodules" that do not secrete T4. Cold nodules are 10 times more likely than hot nodules to become cancerous (to spread to other body parts to form metastases). Hot nodules are very rarely cancerous.

Toxic nodules are an example of a principle we have seen before. The thyroid cells are controlled by the grow-and-secrete feedback circuit in which a signal, TSH, makes the cells both secrete more hormone and to proliferate. The circuit is analogous to the beta-cell circuit (Figure 3.5.13).

This circuit has a fragility as we have seen: Mis-sensing mutant cells that "think" there is too much signal can divide and form a nodule that secretes too much hormone.

This is exactly what happens in a toxic thyroid nodule. The nodule cells are all copies (a clone) of an original mutant cell with a mutation, usually in the TSH receptor, that makes it more sensitive to TSH. There are at least 50 such known mutations in the TSH receptor gene that cause toxic nodules. The mutant cell does what it thinks is right – it is misinformed by the mutant receptor to think that TSH is high so that it must divide and secrete more thyroid hormone. The mutant cell proliferates and over many years grows to a toxic nodule that causes hyperthyroidism.

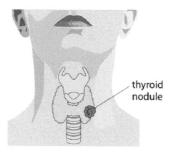

FIGURE 3.5.12 A thyroid nodule is a growth in which a mutant thyrocyte cell proliferates excessively.

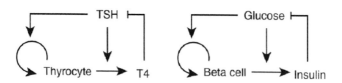

FIGURE 3.5.13 The thyrocytes are regulated by a secrete-and-grow circuit analogous to beta cells.

Often the treatment of toxic nodules is their surgical removal.

It is interesting to consider why the thyroid does not have a biphasic mutant resistance mechanism like beta cells, where cells that sense too much glucose kill themselves. Recall that this effect, called glucotoxicity, gives strong mis-sensing mutants a selective disadvantage.

Why does the thyroid lack a biphasic effect, something we might call TSH toxicity? It would be nice if a cell that senses high TSH would kill itself to avoid toxic nodules. However, unlike beta cells, the signal in the thyroid axis, TSH, varies over a 1000-fold range to compensate for physiological variation, such as iodine levels in the nutrition which can vary over at least a 100-fold range. If there was TSH toxicity, thyroid cells would kill themselves when iodine is low, which would be lethal. Blood glucose has a much smaller range of variation and is thus a good candidate for biphasic regulation. The thyroid is thus prevented from using a TSH toxicity due to its need for extensive compensation.

WHY ARE THERE AUTOIMMUNE DISEASES OF HORMONE GLANDS?

Why is the thyroid attacked by the body? Why risk 2% of the population with a potentially lethal disease? Hashoimoto's thyroiditis was deadly before medicine turned it into a curable disease. The same question for beta cells, where type-1 diabetes is an autoimmune attack on beta cells that was a death sentence to about 1% of children before the advent of insulin treatment.

Also, why these particular cell types and not others? For example, right next to the beta cells are alpha cells that produce glucagon. Why is there no autoimmune disease that attacks alpha cells?

If you want to consider some answers, read on.

EXERCISES

Exercise 3.5.1: A model for compensation in the thyroid axis

a. Write equations for the thyroid axis, identical to the HPA axis except that x_3 inhibits P growth instead of x_1 enhancing P growth (Korem Kohanim et al. 2022).

b. What is the steady state concentration of the three hormones? Compare this to the *HPA* axis.

c. When iodine is lacking, the thyroid is less effective at making thyroid hormone, x_3. As a result, there is compensation. Model low iodine by reducing the x_3 synthesis rate parameter q_3. How do the hormones x_1, x_2, and x_3 change as a function of q_3? How do the gland sizes change? Why is there often an enlarged thyroid gland in regions of low iodine, a condition called Goiter?

d. Why is the hormone x_2, called TSH, considered an excellent clinical blood test for thyroid problems?

Exercises 3.5.2: Graves' disease

Graves' disease is an autoimmune disease that causes hyperthyroidism in about 1% of the population, usually around middle age. In Graves' disease, the body produces antibodies that activate the TSH receptor, mimicking TSH. As a result, the thyroid produces more thyroid hormones.

a. Use the thyroid model of Exercise 3.5.1 to model Graves' disease, by changing x_2, the hormone TSH, to $x_2 + Ab$, where Ab are the activating auto-antibodies.

b. Plot the levels of the hormones and size of the glands as a function of Ab.

c. Explain the transition from subclinical hyperthyroidism to clinical hyperthyroidism. What happens to the pituitary cell mass P at the transition point (Korem Kohanim et al. 2022)?

FURTHER READING

Chaker, L., S. Razvi, I. M. Bensenor, F. Azizi, E. N. Pearce, and R. P. Peeters. 2022. "Hypothyroidism." *Nature Reviews Disease Primers* 8 (1): 1–17. https://doi.org/10.1038/s41572-022-00357-7.

Chatzitomaris, A., R. Hoermann, J. E. Midgley, S. Hering, A. Urban, B. Dietrich, A. Abood, H. H. Klein, and J. W. Dietrich. 2017. "Thyroid Allostasis–Adaptive Responses of Thyrotropic Feedback Control to Conditions of Strain, Stress, and Developmental Programming." *Frontiers in Endocrinology* 8: 163.

Hadlow, N. C., K. M. Rothacker, R. Wardrop, S. J. Brown, E. M. Lim, and J. P. Walsh. 2013. "The Relationship between TSH and Free T4 in a Large Population Is Complex and Nonlinear and Differs by Age and Sex." *The Journal of Clinical Endocrinology & Metabolism* 98 (7): 2936–43. https://doi.org/10.1210/jc.2012-4223.

Korem Kohanim, Y., T. Milo, M. Raz, O. Karin, A. Bar, A. Mayo, N. M. Cohen, Y. Toledano, and U. Alon. 2022. "Dynamics of Thyroid Diseases and Thyroid-Axis Gland Masses." *Molecular Systems Biology* 18 (8): e10919. https://doi.org/10.15252/msb.202210919.

II

Immune Circuits

CHAPTER 4

Autoimmune Diseases as a Fragility of Mutant Surveillance

This chapter is in memory of Nir Friedman.

INTRODUCTION

Hormone glands are prime targets for autoimmune attack. We saw how in type-1 diabetes (T1D) the immune system, namely T cells, kills beta cells. In Hashimoto's thyroiditis, T cells kill the thyroid cells. In both cases, important hormones are lost, insulin and thyroid hormone, and this loss can be fatal unless treated by lifelong supplement of the missing hormones. The diseases often occur at a young age and are very common, 1% of the population has T1D and 2%, primarily female, has Hashimoto's thyroiditis. Why did evolution fail to eradicate these diseases? Why does the body attack itself?

Another question is why the immune system targets these specific cell types and not others? Figure 4.1 shows how some glands get common autoimmune diseases while other glands very rarely get these diseases. Are there rules for which organ gets attacked and which is spared?

In this chapter, we discuss from first principles why endocrine autoimmune diseases arise. Here is the main idea: as we saw, these glands have a circuit that is essential for size control and robustness; this circuit is, however, fragile to mis-sensing mutants that secrete too much hormone and proliferate. To avoid mutant take-over, we will explore the hypothesis that the body uses T cells to remove the mutants.

These autoimmune T cells thus serve an essential role in healthy individuals. But in some individuals, they go rogue and cause autoimmune disease. Thus, there is a tradeoff between risk of autoimmune disease and risk of diseases of hyper-secreting mutant expansion. Different tissues choose among these two evils according to the evolutionary costs and benefits, and in this chapter, we will deduce rules for which tissues get autoimmune diseases versus mutant-expansion diseases.

DOI: 10.1201/9781003356929-9

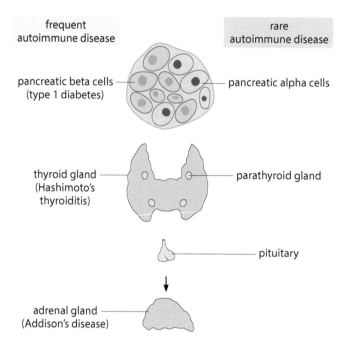

FIGURE 4.1 Some cell types get prevalent autoimmune diseases which specifically kill them, whereas other cell types do not.

TYPE-1 DIABETES IS A DISEASE IN WHICH THE IMMUNE SYSTEM KILLS BETA CELLS

In type-1 diabetes, beta cells are attacked and killed by the body's own immune system. When enough beta cells are killed, insulin levels in the blood are insufficient and glucose can't get into the cells from the blood. The cells starve, and switch to metabolizing fats, acidifying the blood below the normal pH of 7.35, which is deadly.

Until the 1920s, T1D was a death sentence for the children who got it. The discovery of insulin 100 years ago by Banting and Best allowed T1D patients to survive by injecting insulin at the proper doses and times, nowadays using an automated pump. But T1D still causes suffering and morbidity and is not easy to control. It is not known how to prevent T1D, causing special concern for people at risk, such as family members of a person with T1D. The fundamental reason that the body attacks specific cells, beta cells in this case, is also not known. As usual in medicine, when the origin is unknown, it is discussed as a combination of genetic and environmental factors.

It is remarkable that T1D is so prevalent and has such a young age of onset (peaking around age 14), because this is a huge evolutionary cost. Natural selection should have eradicated this disease, especially the self-killing immune cells. The fact that these cells are not eliminated raises the possibility that the disease represents the dark side of an important physiological process.

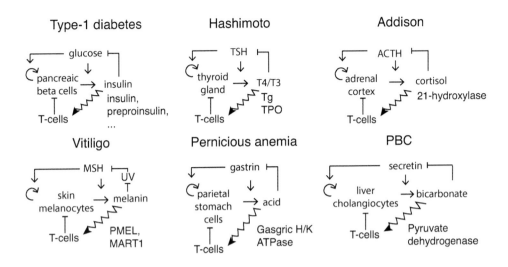

FIGURE 4.2   Cell types with autoimmune diseases share the secrete-and-grow circuit motif.

MANY ENDOCRINE ORGANS HAVE ORGAN-SPECIFIC AUTOIMMUNE DISEASE

T1D and Hashimoto's are just two of many autoimmune diseases. Autoimmune diseases are classified into systemic diseases that attack many organs (like lupus and rheumatoid arthritis), and **cell-type-specific diseases** such as T1D. Here we focus on the latter. These diseases happen primarily in hormone-secreting organs (endocrine organs) or other secretory organs. There is a range of such diseases. Relatively common diseases with a prevalence of 0.01%–0.1% are Addison's disease of the adrenal cortex, vitiligo of the skin melanocytes, and gastritis of the stomach parietal cells called pernicious anemia (Figure 4.2). The origin of all these diseases is currently unknown: they are said to be a combination of genetic and environmental factors.

Equally puzzling is the fact that some endocrine organs virtually never get autoimmune diseases (Figure 4.1). These include the pituitary, alpha cells that secrete glucagon and parathyroid cells that secrete a hormone that controls calcium (PTH). We will try to understand why in this chapter.

All these organ-specific diseases are due to white blood cells called *T* cells attacking the specific cell type that secretes the hormone. Antibodies from *B* cells also participate in the carnage. Why does our immune system attack our own body?

The immune system is designed to protect us against pathogens like bacteria and viruses, and to eliminate cancer cells. The key players are white blood cells called *T* cells, so let's learn more about them.

EXECUTIVE SUMMARY OF *T*-CELL BIOLOGY

The *T* cells normally attack virus-infected cells, not healthy cells of the body. How do they identify which cells are infected? Cells in the body display small pieces of the proteins they make on "identity cards" on their surface, protein complexes called MHCs. Cells infected

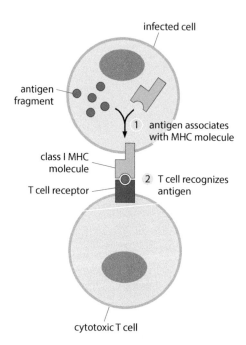

FIGURE 4.3   A cytotoxic T cell recognizes and kills a virus-infected cell when the infected cell presents a viral antigen fragment in an MHC complex on its surface that binds the T-cell receptor.

by virus therefore display pieces of the virus proteins on some of the MHCs. For example, COVID-19 virus forces the cell to make the viral spike protein, and cells display pieces of spike in MHC on their surface.

The MHCs are scanned by T cells. Each T cell has a special receptor called the T-cell receptor or TCR. Each TCR can sense specific pieces of protein, called **antigens**, when presented on an MHC.

T cells that sense normal body proteins are dangerous because they might kill healthy cells. Therefore, they are usually eliminated in an "education" process called tolerance. The remaining T cells respond to foreign proteins and can kill virus infected cells (Figure 4.3). When a T cell recognizes a viral antigen, like the COVID-19 spike protein, it kills the cell by injecting poison and setting off suicide pathways. The T cells also activate B cells to make **antibodies** against the same targets, and the antibodies mark the virus for destruction. The COVID-19 vaccine works by making muscle cells produce spike protein, inducing T cells and antibodies that can kill the virus.

Unfortunately, in T1D, T cells recognize insulin precursors as antigens and kill beta cells. In Hashimoto's thyroiditis, T cells attack healthy thyroid cells. The antigens recognized by T cells are pieces of the Tg and TPO proteins that are the final stages in production of thyroid hormone. The T cells also activate B cells to make antibodies against Tg and TPO. Hashimoto's thyroiditis is identified in the clinic by blood tests for anti-Tg and anti-TPO antibodies.

AUTOIMMUNE *T* CELLS ARE THOUGHT TO BE ERRORS

The most common hypothesis until recently is that autoimmune diseases are mistakes, failures of **tolerance** mechanisms that eliminate dangerous *T* cells. The purpose of tolerance is to eliminate *T* cells that detect self-proteins. This is done when *T* cells are made in the thymus. Their receptors are compared to a vast library of self-proteins in the thymus, and self-reactive cells are eliminated or turned into regulatory *T* cells T_{regs} which act to reduce *T*-cell activity.

The regulatory *T* cells are important elements to control against autoimmunity; rare congenital mutations in T_{regs} often lead to autoimmune attack of multiple endocrine organs. So do mutations that destroy selection in the thymus (AIRE mutations).

T cells that escape elimination in the thymus can still be eliminated or suppressed in the rest of the body when they are over-activated by self-proteins or activated out of context in the periphery.

Still, these processes do not eliminate all self-reactive *T* cells. Research over decades has shown that there are self-reactive *T* cells in all healthy people. How these self-reactive *T* cells sit quietly is not fully understood (Semana et al. 1999; Madi et al. 2014; Yu et al. 2015; Culina et al. 2018; Li et al. 2022; Hs et al. 2021).

Thus, mainstream thought is that self-reactive *T* cells are errors in the tolerance mechanisms. A different line of thought in immunology is that **self-reactive *T* cells play maintenance roles in the body** (Kracht et al. 2016; Schwartz and Cohen 2000; Schwartz and Raposo 2014). As always in this book, we will go with this line of thought – that what appears to be an error or arbitrary detail actually has a functional role. This outlook gives you the chance to make discoveries that you might otherwise miss.

WE EXPLORE THE IDEA THAT *T* CELLS CAN HELP TO REMOVE HYPER-SECRETING MUTANTS

The organs that get organ-specific autoimmune disease share the same circuit motif as beta cells and thyroid cells. In this secrete-and-grow circuit, which we have explored at length in the previous chapters, a signal causes the cells both to secrete a hormone and to proliferate (Figure 4.2). All of these tissues are thus sensitive to mutants that mis-sense the signal. Such mis-sensing mutants can expand and secrete too much hormone, causing loss of homeostasis.

Such mutants are well known clinically. As we saw in the previous chapter, thyroid cells with mutations in the receptor for their signal (TSH) grow into nodules that secrete too much thyroid hormone. These toxic nodules cause hyper-thyroidism which can be lethal. Incidentally, these nodules are not cancerous – unlike cancer, they don't give rise to new growths in other tissues called metastasis. They are instead adenomas which behave like normal thyroid cells, except that the mutant cells "think" there is too much signal. Similarly, certain mutations in beta cells make them think there is too much glucose as described in Chapter 2.

These mutant cells are inevitable. An organ like the thyroid weighs 20 g and has about 10^{10} cells. It thus takes 10^{10} cell divisions to make it. Since mutation rate is about $10^{-9}/$

base-pair/division, each possible point mutation will be found in about 10 thyroid cells. It is known that at least 50 such mutations cause hyper-sensing and hyper-secretion leading to toxic thyroid nodules. Thus, every person should develop at least $10 \times 50 = 500$ toxic thyroid nodules secreting thyroid hormone – which would be lethal. Similarly, the 10^9 beta cells are sure to get enough insulin hyper-secreting mutants to kill the person from hypoglycemia. Thus, just to have functional endocrine organs requires removal of mutants.

In Chapter 2, we saw a **biphasic mechanism** for removing such mutants in beta cells: glucotoxicity. Glucotoxicity causes mutants that "think" glucose is too high to kill themselves. We noted that this mechanism still leaves the range of mild mis-sensing mutants, between the two fixed points (hatched region in Figure 4.4). Other organs, like the thyroid, do not have a mechanism like glucotoxicity at all. There is no TSH toxicity, because TSH needs to vary over a 1000-fold range in normal physiology, such as when iodine levels in nutrition change. Thus, we need another mechanism to remove mutants.

In this chapter, we consider the idea that T cells can help to remove mis-sensing mutants (Korem et al. 2020). To eliminate these mutants, we need a surveillance mechanism, which we will call Autoimmune Surveillance of Hyper-secreting Mutants (ASHM) (Figures 4.5 and 4.6) or **autoimmune surveillance** for short.

This is a theory we developed with Yael Korem Kohanim during her PhD and systems immunologist Nir Friedman (Korem Kohanim et al. 2020). This chapter is in memory of Nir Friedman, who passed away in 2020. A noble, gentle, and clear thinker.

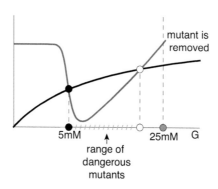

FIGURE 4.4   The biphasic mechanism makes mutant cells that strongly hyper-sense the input signal remove themselves.

FIGURE 4.5   Overview of the autoimmune surveillance theory in which auto-reactive T cells remove hyper-sensing mutant cells.

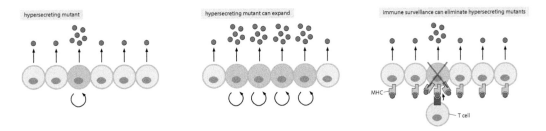

FIGURE 4.6 Hyper-sensing mutant cells hyper-secrete the hormone and proliferate, expanding into a clone of mutant cells that leads to excessive hormone levels. These cells can be removed by the autoreactive *T* cells in the autoimmune surveillance mechanism.

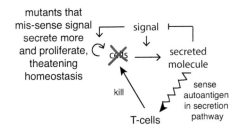

FIGURE 4.7 In the autoimmune surveillance theory, *T* cells that sense the autoantigen preferentially kill cells with more antigen then their neighbors, eliminating hyper-secreting mutant cells.

Autoimmune surveillance needs to detect the hyper-secreting cells in order to eliminate them. Thus, the antigens it detects must be in the production pathway of the hormone. In this way, cells that make more hormones than their neighbors will present more of these antigens on their surface (Figure 4.7). *T* cells can then detect the hyper-secreting mutant cells based on their large number of presented hormone-related antigens.

The best idea is to detect antigens at the very end of the production pathway of the hormone. That design can capture many different mutations that lead to hyper-secretion. Indeed, the antigens in T1D (called **autoantigens**) are all pieces of proteins in the insulin secretion pathway. For example, a major antigen is pre-proinsulin, the very last protein in the production pipeline, which is cleaved to make insulin. Other major T1D auto-antigens are also proteins in the secretion pathway of insulin.

Antigens from proteins at the end of the hormone production pathway are found in all the organ-specific autoimmune diseases: the autoantigen in Hashimoto's thyroiditis is the protein cleaved to make thyroid hormone (thyroglobulin *Tg*, the analog of pre-proinsulin), or the key enzyme that modifies this protein to make the hormone (TPO). In Addison's disease, the autoantigen is the key enzyme that synthesizes cortisol (21-hydroxylase). Other examples are shown in Figure 4.2 and in Table 4.1.

The *T* cells that recognize pre-proinsulin and other auto-antigens are found in the *T*-cell repertoire shared by all people, called the **public *T*-cell repertoire** (Korem et al. 2020; Madi et al. 2014).

TABLE 4.1 The Autoantigens in Many Cell-Type-Specific Autoimmune Diseases Are Fragments of Proteins That Play a Role in the Production and Secretion of the Hormone or Metabolite Produced by the Cell

Autoimmune Disease	Auto-Antigens	Role of Auto-Antigen
Type 1 diabetes	Insulin, preproinsulin, PTPRN, PTPRN2, islet cell antigen-69, ZnT8, GAD65	Insulin synthesis, storage, and secretion
Hashimoto's thyroiditis	Thyroid peroxidase, thyroglobulin	Thyroid hormone biosynthesis
Addison's disease	21-hydroxylase	Cortisol/aldosterone biosynthesis
Vitiligo	PMEL, MART1, tyrosinase, tyrosinase related proteins 1 and 2	Melanin synthesis and storage
Autoimmune gastritis	Gastric H/K ATPase	Acid production
Primary biliary cirrhosis	PDC-E2 pyruvate/oxo-glutarate dehydrogenase	Bicarbonate production

T CELLS CAN TELL THE DIFFERENCE IN ANTIGEN BETWEEN NEIGHBORING CELLS

For immune surveillance to work, the killer *T* cells need to tell which cell makes more antigen than its neighbors, so that they can preferentially kill hyper-secreting cells.

Such differential sensitivity is indeed a feature of *T* cells. Experiments tested the relation between the amount of an antigen that a cell presents and the probability that it is killed by a *T* cell that recognizes that antigen. The probability of killing is often an S-shaped function of the number of MHCs on the cell surface that present the antigen (Figure 4.8) (Pettmann et al. 2021; Martin-Blanco et al. 2018; Halle et al. 2016).

For *T* cells that bind their antigen very strongly, a single antigen presented on a cell is enough. Thus, autoimmune surveillance cannot operate with very strong binding *T* cells, because they would not be able to discriminate between hyper-secreting cells and normal cells.

Many *T* cells, however, have only moderate binding to their antigens. For moderate binding, the killing rate $h(a)$ goes approximately as a power law of antigen level a

$$(4.1) \qquad\qquad h(a) \approx c\, a^n,$$

with a large exponent $n = 3$–5, signifying a steep relationship (Figure 4.8). We saw such steep relationships in previous chapters, for example in glucose sensing by beta cells. Steep relationships in biology are often caused by mechanisms of "**cooperativity**," for example

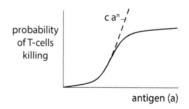

FIGURE 4.8 *T* cells with intermediate affinity kill target cells according to a sigmoidally rising function of presented antigen. Dashed line is a power law approximation of a Hill function, with steepness determined by n.

when multiple T cell receptors in the same T cell cluster together and cooperate to make each other more active.

Another important property of the immune system is that it can adapt to a background level of antigen and only respond to temporal changes in antigen. This adaptation to background is provided, in part, by **regulatory T cells**. T_{regs} provide an incoherent feedforward loop circuit that has the capacity to adapt to a constant input signal (antigen level) and to respond to exponentially increasing antigen threat (Sontag 2017). This circuit is explored in Solved Exercise 4.2. Other mechanisms exist to help the T cells adapt, such as molecular "switches" on the T cell that make them less active if they kill too often, called immune checkpoints. The result of this adaptation is that killing rate goes according to the ratio of antigen relative to the mean antigen presented by all cells, a/<a>, so that

(4.2)
$$h(a) = c\left(\frac{a}{\langle a \rangle}\right)^n$$

This killing function therefore has two parameters: the rate c and the cooperativity n.

The relative sensing explains why the T cells don't severely attack an organ if it simply starts to produce more hormone. For example, when beta cells start making more insulin due to a change in diet or insulin resistance. More hormone is made in all the cells of the organ, more antigen is presented, T_{regs} level rise and compensate by inhibiting the effector T cells. The immune system thus adjusts to precisely cancel out the rise in antigen. It remains sensitive to individual cells that make more antigen than their neighbors.

AUTOIMMUNE SURVEILLANCE CAN ELIMINATE ANY MUTANT, AND CAN DO SO WITH A LOW KILLING RATE

To work well, autoimmune surveillance needs to eliminate any possible hyper-sensing mutant, and to do so without killing too many healthy cells. To understand how this might work, we analyze a mathematical model for autoimmune surveillance in Solved Exercise 4.1.

The main conclusion from the model is that autoimmune surveillance can work effectively, and silently - it can eliminate mutants while only rarely killing normal cells. The model also shows that autoimmune surveillance is an **"evolutionary stable strategy"** (ESS): a mechanism in which no mis-sensing mutant cell can invade and outgrow a large population of normal cells. No matter what is the "perceived glucose" as in Figure 4.9, the mutant cell has a growth disadvantage compared to normal cells. Autoimmune surveillance thus effectively eliminates sporadic mutant cells.

Of course, if all the cells are mutant, as in rare mutations that occur in the fertilized egg and are present in all cells of the body, autoimmune surveillance cannot detect cheater cells – it relies on differences between cells. This occurs in very rare conditions of congenital hyperinsulinemia, in which all beta cells are hyper-secreting mutants and babies are born with low glucose, sometimes necessitating removal of the pancreas.

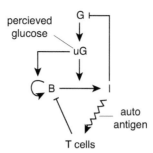

FIGURE 4.9   Circuit for autoimmune surveillance in the beta cell-insulin-glucose system. A mutant cell senses u times more glucose than really exists.

The model also estimates the collateral killing of normal cells by autoimmune surveillance. This off-target killing is prevented by a steep T cell recognition curve, described in Figure 4.9 by high Hill coefficients n in Eq. 4.2. Autoimmune surveillance kills much fewer cells than the natural turnover of the tissue when n is large, $n=2$ or above. Thus, only a small part of the removal of beta cells is due to autoimmune surveillance, and the rest to natural turnover.

SURVEILLANCE CAN DESCEND TO AUTOIMMUNE DISEASE IN SEVERAL WAYS

Most people don't get autoimmune disease, suggesting that this surveillance mechanism works well. But a small fraction of people unfortunately get autoimmune disease. The risk has a sizable genetic component. But genetics is not all, there is also a stochastic component – even identical twins have only about a 50% concordance in terms of getting autoimmune disease.

How might autoimmune surveillance fail and descend to autoimmune disease? We don't know for sure. One theory is that a viral or bacterial infection damages the tissue, causing release of self-antigen. The infection grows exponentially and provides an inflammation danger signal. The T cells see an exponentially rising amount of self-antigen in the context of inflammation. It concludes wrongly that the self-antigen, such as pre-proinsulin, is actually of viral origin.

Genetic factors come into play, such as **MHC variants**. The MHC genes are polymorphic, meaning that people carry different variants. The MHC variants that pose a risk for autoimmune disease include HLA-DR3,4. They encode class-2 MHCs, which are used by antigen-presenting cells. Such class-2 MHCs present antigens to other immune cells and play a role in activating B cells and in regulatory T cell function. The high-risk variants may help to set off the heavy guns, the **B cells** that produce antibodies against the antigen. The antibodies coat the beta cells and act as a "kill me" signal. Immune cells then attack the beta cells aggressively, thinking that they have viruses inside them (Figure 4.10).

Another possibility, raised by experiments in mouse, is that these high-risk MHC variants cause T_{regs} to kill themselves at high antigen levels (a kind of biphasic response). When T_{regs} are gone, there is less inhibition on killer T cells, unleashing a large autoimmune response.

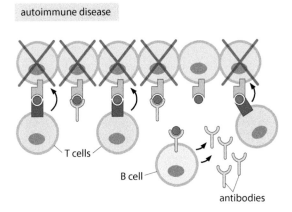

FIGURE 4.10 When autoimmune surveillance falsely triggers antibody-producing *B* cells, enhanced and indiscriminate damage to the tissue can result. This is full-fledged autoimmune disease.

Note that these genetic risk factors are relatively common. They were probably selected in the past because they played a beneficial role against a dangerous pathogen. By enhancing *B*-cell activation or reducing T_{regs} inhibition, such variants set off a powerful immune response and helped to better fight the pathogen. This may explain why these genetic risk factor variants are present in a sizable fraction of the population, about 30%, quite consistently across the world population.

In a normal response to pathogens, the antibody and *T* cell responses stop when the pathogen is eliminated, and the foreign antigen is gone. In the aftermath of infection, *T* cells even kill each other in a process called fratricide. But in autoimmune disease, cells of the targeted tissue are attacked and killed relentlessly. The killing releases more self-antigen, activating more immune cells, making a **vicious cycle**. Long-lasting **memory** *B* cells and *T* cells are formed which are easily triggered by the antigen. When about 90% of the beta cells or thyroid cells are killed, hormone production drops so low that clinical symptoms set in.

Whatever the precise route to autoimmune disease, the presence of auto-reactive *T* cells provides a potential fragility to self-attack.

ENDOCRINE TISSUES THAT RARELY GET AUTOIMMUNE DISEASE ARE PRONE TO DISEASES OF MUTANT EXPANSION

We now turn to the question of why certain organs are attacked and not others (Figure 4.1). The autoimmune surveillance theory predicts a tradeoff: if there is little or no surveillance in a tissue, it should get no autoimmune disease. However, it should get diseases of mutant expansion, especially at old ages when mutant cells have had enough time to grow into a large nodule called an adenoma.

We can test this prediction by looking at endocrine cells and organs that very rarely have autoimmune diseases – less than $10^{-5} - 10^{-6}$ lifetime prevalence. These organs include the parathyroid (PT) gland, a tiny gland that sits on top of the thyroid (Figure 4.1). Its job is to secrete the hormone PTH in order to control free blood calcium. PTH helps dissolve bone, which is made of calcium phosphate, and to regulate calcium balance from the gut and kidney, in order to increase blood calcium.

The lack of autoimmune disease in this gland suggests that it has no autoimmune surveillance or perhaps a weak version. **This predicts that the gland is prone to expansion of hyper-secreting mutants.** Indeed, such mutant expansion is a common disease with the long name *primary hyperparathyroidism*. It afflicts about 1/50 women after menopause. A hyper-secreting mutant cell grows exponentially and becomes an adenoma, secreting too much hormone and pushing calcium levels up. The excessive calcium comes at the expense of bones, and the symptoms include loss of bone mass and neuronal problems. Treatment sometimes requires surgically removing the adenoma.

The PT gland has a secrete-and-grow circuit that is sensitive to take-over by mis-sensing mutants. The circuit is essentially the same as in the other glands, except for a sign reversal. In this circuit (Figure 4.11), the signal, calcium, inhibits both the proliferation of PT cells and the secretion of PTH. A mutant cell that mis-senses normal calcium as too little calcium (hypo-sensing mutant) is the culprit: such a mutant expands and hyper-secretes PTH, and if it grows to a sizable adenoma, it leads to excessive calcium. This circuit also has biphasic mutant protection (low and high calcium kills PT cells). But intermediate mis-sensing mutants are still dangerous. It is precisely such mild mis-sensing mutations in the calcium receptor that cause the adenomas in the parathyroid gland.[1]

Thus, a tradeoff seems to exist between two evils: autoimmune disease and diseases of hyper-secreting mutant expansion.

Another example of this mutant/autoimmunity tradeoff occurs in the *HPA* axis. Recall the two glands, the pituitary and the adrenal, A and P. Each has a version of the secrete-and-grow circuit motif that is fragile to mis-sensing mutants, as we saw in Chapter 3. Mutants in the pituitary that hyper-sense hormone x_1, for example, grow into nodules that hyper-secrete hormone x_2 (ACTH), making the adrenal secrete too much cortisol (Figure 4.12). This is known as **Cushing's syndrome**, with depression, hypertension, muscle wasting, and fat distribution in the face and abdomen. The same disorder can be caused by adrenal mutants that hyper-sense x_2. However, 90% of Cushing's syndrome is caused by mutants in P, not in A. This is surprising because the number of cells in P is smaller by a factor of 100 than in A (relevant cells in the adrenal total about 10 g, in the pituitary about 0.1 g).

Our theory predicts then that the adrenal A is protected from mutants by autoimmune surveillance, and hence should have autoimmune disease. Indeed, the adrenal is destroyed

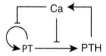

FIGURE 4.11 The parathyroid hormone system shows the secrete-and-grow circuit motif and has prevalent expansions of hyper-secreting mutant cells that lead to excessive blood calcium.

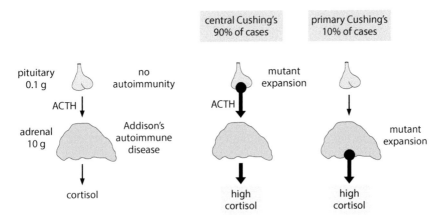

~1000-fold higher probability of hypersecreting mutant per cell in pituitary compared to adrenal

FIGURE 4.12   The HPA axis shows an example of the inverse relation of autoimmune disease and hyper-secreting mutant-expansion disease. The adrenal gets an autoimmune disease whereas the pituitary almost never does. Conversely, the chance per cell of a hyper-secreting mutant expansion is about 1000 times larger in the pituitary.

by *T* cells in an autoimmune disease called Addison's disease. The pituitary virtually never gets such an autoimmune disease (unless caused by certain checkpoint inhibitor drugs). It seems to have little autoimmune surveillance, perhaps because it is an immune-privileged site. However, as mentioned, it shows relatively frequent mutant-expansion diseases – the most common form of Cushing's syndrome called central Cushing's (Figure 4.12).

Similar pituitary mutant expansion diseases plague other *HP*-axes. Pituitary mutant cells in the growth axis account for acromegaly and gigantism, and in other pituitary pathways to disease of hyper-thyroidism. Again, like the adrenal, the thyroid is prone to autoimmunity, whereas its pituitary controller cells (TSH-secreting thyrotroph cells) are prone to mutant expansion.

What rules might determine if a tissue gets autoimmune disease or diseases of mutant expansion? One possibility is based on the evolutionary cost of these diseases: it pays to set things up so that the less severe disease occurs (Figure 4.13). In beta cells, a hyper-secreting mutant expansion is lethal, because it causes low glucose. Thus, it makes sense to have strong autoimmune surveillance, which can sacrifice some of the population to T1D, but can save a higher fraction of the population from lethal mutant expansion disease.

In the PT gland, in contrast, high levels of calcium caused by mutant expansion are bad but not lethal. This gland has a biphasic mechanism to protect against strong hyper-secreting mutants. The mild mutants that take over cause high calcium, but below the lethal calcium level of roughly four times the normal level. But a reduction in PTH, as would be caused by autoimmune destruction of the PT gland, can push calcium down to lethal levels: even a 20% reduction is lethal (below 0.9 mM compared to the normal level of 1.1 mM). Thus, it makes sense that autoimmune surveillance does not evolve in the PT gland, to avoid the risk of low calcium, at the price of a less severe mutant expansion disease with a late age of onset.

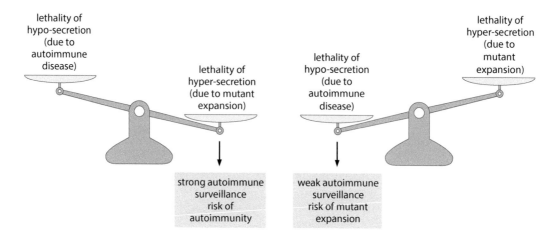

FIGURE 4.13 One explanation for the prevalence of autoimmune versus mutant expansion disease is natural selection of immune surveillance levels so that the more lethal disease is rarer.

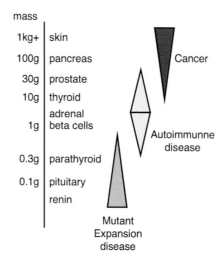

FIGURE 4.14 The major disease class of a secretory cell type depends on the total number of cells of that type in the body. Hypersecreting mutant expansion tends to occur below 10^9 cells, autoimmune diseases in the range of 10^9–10^{10} cells, and cancer above about 10^{10} cells.

Perhaps the simplest explanation for which organ gets which disease stems from the number of mutations in the organ. The smaller the organ, the fewer cell divisions are needed to make it, and the fewer cell divisions occur over life. The fewer the mutations, the less the need for autoimmune surveillance. The cutoff seems to be at a mass of about 1 g, which is about 10^9 cells. Above 1 g are endocrine organs with autoimmune diseases: beta cells (1 g), adrenal (10 g), thyroid (10 g), as shown in Figure 4.14. Below 1 g are glands with mutant expansion and very rare autoimmunity: pituitary cells (0.1 g), parathyroid (0.3 g), renin secreting cells (0.1 g).

At above 10–30 g, the disease spectrum shifts again, with less autoimmunity and more cancer (prostate 30 g, pancreas 100 g, skin several kg), Figure 4.14. This is because there are so many cell divisions that mutations become too frequent. The required levels of autoimmune surveillance would be too high to avoid autoimmune disease. Organs thus change strategy to stem-cell-based production, in which a single stem-cell division is amplified to make thousands of cells by transit amplifying cells. This reduces the number of mutations that remain in the stem cells of the tissue. But stem cells are more prone to cancer, being cells with high proliferative potential. In the transition zone of 10–30 g, one sees both cancer and autoimmunity, as in the thyroid and prostate (Figure 4.14)

I like the prospect of such rules for diseases, pointing towards the periodic table of diseases that will be our closing chapter. The table can predict diseases based on first principles. We will expand on this theme later on. But now we turn to understand another common pathology of the immune response – inflammation and excessive scarring known as fibrosis.

EXERCISES

Solved Exercise 4.1: Develop and analyze a model for autoimmune surveillance of hyper-secreting mutants. Show the conditions where autoimmune surveillance cannot be invaded by any mutant cell, known as an evolutionarily stable strategy.

Since we are dealing with secrete-and-grow circuits, we use beta cells as an example. We begin with the growth equation for beta-cell mass B from Chapter 2, whose growth rate is controlled by glucose G:

$$(4.3) \qquad \frac{dB}{dt} = \mu(G)B$$

We first consider healthy conditions, where glucose is near the stable fixed-point $G = G_0 = 5$ mM. Recall that the beta-cell growth rate is zero at G_0, so that biomass growth equals removal and beta-cell mass is at steady state (Figure 4.15).

Near G_0, we can approximate the net growth rate as a line with slope denoted μ_0, so that $\mu = \mu_0(G - G_0)$ (gray line in Figure 4.15). This linear approximation is not essential but makes the math easier and is sufficiently accurate for our purposes. Thus

$$(4.4) \qquad \frac{dB}{dt} = \mu_0(G - G_0)B$$

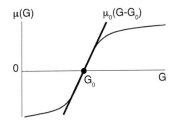

FIGURE 4.15   A linear approximation for the net growth rate near the fixed point.

Now let's add autoimmune surveillance, in which beta cells are killed by T cells. We begin with the case in which there are only non-mutant beta cells called **wild-type** cells. Each cell presents a copies of antigen on its surface. The antigens are in the secretion pathway and hence proportional to the insulin production rate per cell. Inserting the killing term from Eq. 4.1 into the growth equation, we find

$$(4.5) \qquad \frac{dB}{dt} = \mu_0(G - G_0)B - c\left(\frac{a}{\langle a \rangle}\right)^n B$$

Since all cells are wild-type cells, $<a> = a$, and the killing term is just equal to $c1^n = c$. Thus, at steady state,

$$(4.6) \qquad \frac{dB}{dt} = 0 = \mu_0(G - G_0) - c$$

whose solution is the steady-state glucose level that is slightly shifted upward from $G0$ due to the effect of beta-cell killing:

$$(4.7) \qquad G_{st} = G_0 + c/\mu 0$$

The higher the killing rate c, the higher the glucose because more beta cells are killed per unit time, and hence less insulin, and thus more glucose. In extreme cases, where c is very large, killing is widespread, and we have very high glucose levels – this is the situation in autoimmune disease like T1D.

Now let's consider a mutant beta cell that mis-senses glucose. It acts as though the true glucose level G is actually uG, where u is the **mis-sensing factor** (Figure 4.16). Such mis-sensing mutants were discussed in Chapter 2. The mutant cells have enhanced insulin secretion and proliferation. In the equation for such a mutant, we need to replace all occurrences of G by uG. The equation for the growth of the mutant population B_m is therefore

$$(4.8) \qquad \frac{dB_m}{dt} = B_m\left[\mu_0(uG - G_0) - c\left(\frac{a_m}{\langle a \rangle}\right)^n\right]$$

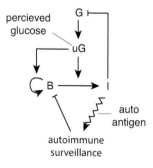

FIGURE 4.16 The autoimmune surveillance circuit, where a mutant cell senses u times more glucose than reality.

where the autoimmune surveillance killing term contains the mutant antigen level a_m. Initially, there is a single mutant cell surrounded by wild-type cells. Since there is only one mutant cell, the average antigen $\langle a \rangle$ is virtually unaffected by the mutant so that $\langle a \rangle$ can be approximated by the wild-type level of antigen. Similarly, glucose level G is virtually unaffected by the mutants as long as they are few.

The antigen level of the mutant is determined by its insulin production rate. Using the insulin equation from Chapter 1, we have $a_m \sim q\, f(uG)$. Using $f(G) \sim G^2$, we find that mutant cell antigen level is $a_m = qu^2\, G^2$. The wild-type antigen level is $\langle a \rangle = qG^2$, because $u=1$ for the wild-type cells. Thus, the mutant killing term $c\left(\dfrac{a_m}{\langle a \rangle}\right)^n$ depends only on the mis-sensing factor u, because the factors q and G^2 cancel out, leaving $\left(u^2\right)^n = cu^{2n}$. Thus:

$$(4.9) \qquad \frac{dB_m}{dt} = B_m \left(\mu_0 (uG - G_0) - cu^{2n} \right) = \mu(u) B_m$$

We conclude that the mutant has a growth rate that is determined by its mis-sensing factor u:

$$(4.10) \qquad \mu(u) = \mu_0\, (uG - G_0) - cu^{2n}$$

In order for autoimmune surveillance to work perfectly, we need the wild-type cells ($u=1$) to have the highest growth rate, higher than all possible mutant cells. This is called an **"evolutionary stable strategy"** (ESS): a mechanism which cannot be invaded by any single mutant. For organ size control, the wild-type cells should have zero net growth rate (proliferation equals removal and thus a steady population size), and all mutants should have negative net growth rate and thus eventually vanish. We therefore need to find a condition such that growth rate $\mu(u)$ is maximal at $u=1$ (Figure 4.17). This occurs when

$$(4.11) \qquad \frac{d\mu(u=1)}{du} = 0 \text{ condition for evolutionary stable strategy}$$

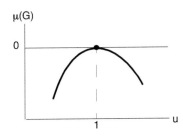

FIGURE 4.17 For an evolutionarily stable strategy – a design that cannot be invaded by any mutant. Net growth rate must be maximal for the non-mutant (wild-type) cells, which sense glucose as it is ($u=1$).

and also that the second derivative is negative to ensure a maximum. Taking the derivative of Eq. 4.8, we find

(4.12)
$$\frac{d\mu}{du} = \mu_0 G - 2ncu^{2n-1} = 0$$

The glucose level G is just the glucose level for the wild-type case (Eq. 4.5) because a single mutant cell can't affect glucose levels much. Plugging in Gst from Eq. 4.5, and $u=1$, we find the condition for an evolutionary stable strategy (ESS):

$$\frac{c}{\mu_0 G_0} = \frac{1}{2n-1}$$

This equation connects the killing rate c, normalized by the natural turnover of beta cells, to the steepness of the killing function n. Interestingly, the higher the T cell cooperativity (steepness) parameter n, the lower the killing rate c that is required for ESS. Since the cooperativity of immune recognition is high ($n \sim 3$–5), killing rate should be small, about 10%–20% of the natural turnover rate parameter $\mu_0 G_0$. Thus, these secret-agent T cells can work subtly and precisely.

Exercise 4.2 An exponential threat detector in the immune system
This exercise explores a circuit that can respond to an exponentially rising threat, such as a virus or bacterium, and ignore signals that do not change with time, such as antigens from self-tissues (Sontag 2017), see Figure 4.18.

a. Consider an antigen $u(t)$ presented by antigen-presenting cells to T cells. This activates effector T cells $T(t)$ that perform the response functions, and also regulatory T cells denoted $R(t)$ that inhibit the effector T cells. Interpret the following equations and their parameters:

$$\frac{dR}{dt} = a\,u - b\,R$$

$$\frac{dT}{dt} = c\frac{u}{R} - d\,T$$

b. Solve the steady state of R and T for a given steady-state level of u. Explain why this steady state does not depend on u.

c. Explain why a step-function change in u from level u_1 to level u_2 leads to a pulse of T activation that returns to its original level (Figure 4.18)

d. Shows that if antigen rises exponentially, $u(t) = u_0\,e^{\alpha t}$, the effector T-cell activity does not return to the original steady-state level. Show that the activation level above their original steady state is proportional to the antigen exponential growth rate parameter α.

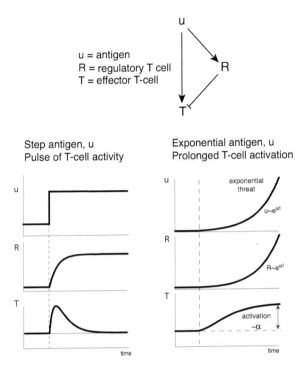

FIGURE 4.18 An exponential threat detector circuit in the immune system. Antigen activates cytotoxic T-cells, T, and also regulatory T-cells, R, that inhibit T. A step increase in antigen causes a pulse of T that adapts back to baseline. Only an exponentially increasing antigen level, as in pathogens or early cancer, causes a prolonged rise in T without adaptation to baseline.

e. What happens when the antigen rises linearly with time $u = \alpha t$?

f. Explain why an exponential threat detector might be useful for the immune system.

Exercise 4.3: Viral dynamics
Consider a model for the concentrations of virus, $u(t)$, effector T cells, $T(t)$, and T_{regs}, $R(t)$:

$$\frac{du}{dt} = (\alpha_0 - c\, T)u$$

$$\frac{dR}{dt} = u - R$$

$$\frac{dT}{dt} = \frac{u}{k+R} - T$$

a. Explain the equations and the parameters k, c, and α_0.

b. Calculate the steady-state solution.

c. Numerically solve the equations for various values of α_0. Use $c = 1$, $k = 1$, $R(0) = T(0) = 0$, and $u(0) = 1$. Explain the meaning of these initial conditions.

d. Assume that when the virus concentration goes below a minimal dose, $u_0 = 0.01$, it is killed by the innate immune system. What is the maximal viral growth rate α_0 for which the virus is killed by the immune system? What happens if α_0 is larger than this value?

Exercise 4.4: Theories for autoimmunity

a. Read about the hypothesis of "molecular mimicry" for autoimmune diseases.

b. Read about the "hygiene hypothesis" for autoimmune diseases.

c. Discuss their pros and cons, and compare to the "surveillance of hyper-secreting mutant" theory discussed in this chapter (200 words).

Exercise 4.5: Bistability in a simple model for autoimmunity
Consider this simple model: The immune system attacks healthy tissue. This releases auto-antigens, making the immune killing stronger, in a cooperative way, with Hill coefficient $n = 2$. The variable is the amount of autoantigen $a(t)$. The autoantigen is removed at rate γ.

a. Explain the equation:

$$\frac{da}{dt} = c \frac{a^n}{k^n + a^n} - \gamma\, a.$$

b. Draw a rate plot showing the fixed points. Consider (graphically) different scenarios (different parameters) with different numbers of fixed points. When is there bistability?

c. Which scenario corresponds to an autoimmune disease? Which corresponds to no autoimmune disease?

d. Suppose that individuals vary in their genetics in a way that affects the parameters of the equation. Does an increase in the parameter c increase the risk for autoimmune disease? Repeat for the parameters k and γ.

NOTE

1 It seems that the parathyroid has at least some autoimmune surveillance because autoimmune disease can be caused by certain drugs. In particular, drugs that enhance immune response against cancer, by blocking immune checkpoints, have parathyroid autoimmunity as a side effect in some patients. Thus, autoimmune surveillance in this organ may be tuned to low levels that prohibit autoimmune disease under normal conditions.

REFERENCES

Culina, Slobodan, Ana Ines Lalanne, Georgia Afonso, Karen Cerosaletti, Sheena Pinto, Guido Sebastiani, Klaudia Kuranda, et al. 2018. "Islet-Reactive CD8+ T Cell Frequencies in the Pancreas, but Not in Blood, Distinguish Type 1 Diabetic Patients from Healthy Donors." *Science Immunology* 3 (20): 4013. https://doi.org/10.1126/SCIIMMUNOL.AAO4013/SUPPL_FILE/AAO4013_SM.PDF.

Halle, Stephan, Kirsten Anja Keyser, Felix Rolf Stahl, Andreas Busche, Anja Marquardt, Xiang Zheng, Melanie Galla, et al. 2016. "In Vivo Killing Capacity of Cytotoxic T Cells Is Limited and Involves Dynamic Interactions and T Cell Cooperativity." *Immunity* 44 (2): 233–45. https://doi.org/10.1016/j.immuni.2016.01.010.

Hs, Wong, K. Park, A. Gola, Ap Baptista, Ch Miller, D. Deep, M. Lou, et al. 2021. "A Local Regulatory T Cell Feedback Circuit Maintains Immune Homeostasis by Pruning Self-Activated T Cells." *Cell* 184 (15). https://doi.org/10.1016/j.cell.2021.05.028.

Korem Kohanim, Yael, Avichai Tendler, Avi Mayo, Nir Friedman, and Uri Alon. 2020. "Endocrine Autoimmune Disease as a Fragility of Immune Surveillance against Hypersecreting Mutants." *Immunity* 52 (5): 872–84.e5. https://doi.org/10.1016/J.IMMUNI.2020.04.022.

Kracht, Maria J. L., Arnaud Zaldumbide, and Bart O. Roep. 2016. "Neoantigens and Microenvironment in Type 1 Diabetes: Lessons from Antitumor Immunity." *Trends in Endocrinology and Metabolism* 27 (6): 353–62. https://doi.org/10.1016/J.TEM.2016.03.013.

Li, Jing, Maxim Zaslavsky, Yapeng Su, Jing Guo, Michael J. Sikora, Vincent van Unen, Asbjørn Christophersen, et al. 2022. "KIR+CD8+ T Cells Suppress Pathogenic T Cells and Are Active in Autoimmune Diseases and COVID-19." *Science* 376 (6590): eabi9591. https://doi.org/10.1126/science.abi9591.

Madi, Asaf, Eric Shifrut, Shlomit Reich-Zeliger, Hilah Gal, Katharine Best, Wilfred Ndifon, Benjamin Chain, Irun R. Cohen, and Nir Friedman. 2014. "T-Cell Receptor Repertoires Share a Restricted Set of Public and Abundant CDR3 Sequences That Are Associated with Self-Related Immunity." *Genome Research* 24 (10): 1603–12. https://doi.org/10.1101/GR.170753.113.

Martin-Blanco, N., R. Blanco, C. Alda-Catalinas, E. R. Bovolenta, C. L. Oeste, E. Palmer, W. W. Schamel, et al. 2018. "A Window of Opportunity for Cooperativity in the T Cell Receptor." *Nature Communications* 9 (1): 2618. https://doi.org/10.1038/s41467-018-05050-6.

Pettmann, Johannes, Anna Huhn, Enas Abu Shah, Mikhail A. Kutuzov, Daniel B. Wilson, Michael L. Dustin, Simon J. Davis, P. Anton van der Merwe, and Omer Dushek. 2021. "The Discriminatory Power of the T Cell Receptor." *eLife* 10: e67092.

Schwartz, Michal, and Irun R. Cohen. 2000. "Autoimmunity Can Benefit Self-Maintenance." *Immunology Today* 21 (6): 265–68. https://doi.org/10.1016/S0167-5699(00)01633-9.

Schwartz, Michal, and Catarina Raposo. 2014. "Protective Autoimmunity: A Unifying Model for the Immune Network Involved in CNS Repair." *Neuroscientist* 20 (4): 343–58. https://doi.org/10.1177/1073858413516799.

Semana, Gilbert, Rudolf Gausling, Richard A. Jackson, and David A. Hafler. 1999. "T Cell Autoreactivity to Proinsulin Epitopes in Diabetic Patients and Healthy Subjects." *Journal of Autoimmunity* 12 (4): 259–67. https://doi.org/10.1006/JAUT.1999.0282.

Sontag, E. D. 2017. "A Dynamic Model of Immune Responses to Antigen Presentation Predicts Different Regions of Tumor or Pathogen Elimination." *Cell Systems* 4(2): 231–241.

Yu, Wong, Ning Jiang, Peter J. R. Ebert, Brian A. Kidd, Sabina Müller, Peder J. Lund, Jeremy Juang, et al. 2015. "Clonal Deletion Prunes but Does Not Eliminate Self-Specific Aβ CD8+ T Lymphocytes." *Immunity* 42 (5): 929–41. https://doi.org/10.1016/J.IMMUNI.2015.05.001.

CHAPTER 5

Inflammation and Fibrosis as a Bistable System

INTRODUCTION

Fibrosis, or excessive scarring, is a medical problem that cuts across medicine. In fibrosis, scar tissue replaces healthy tissue and the organ loses function. Fibrosis occurs in the liver, lung, kidney, heart, and other organs, and is a major contributor to age-related diseases. It is always preceded by periods of intense inflammation. There is currently no cure for progressive fibrotic diseases other than organ transplant.

In this chapter, we will understand the essence of inflammation and fibrosis. Our basic question is how a single biological process, tissue repair, can lead to two very different results: healing or fibrosis. We will use this understanding to consider potential avenues for therapy to prevent fibrosis and even to reverse it.

We will use two of our laws in this chapter, all cells come from cells, and biological processes saturate. The third law, cells mutate, will star again in the next chapters.

INJURY LEADS TO INFLAMMATION, WHICH GOES TO EITHER HEALING OR FIBROSIS

As we know from our childhood injuries, the injured spot gets red, swollen, hot, and painful – this is **inflammation**. The wound develops a scar over a few days. The scar usually vanishes after a couple of weeks, and the tissue is perfectly healed. But sometimes we get permanent scars that last a lifetime. These scars are examples of **fibrosis**.

There is a universal sequence across tissues:

$$injury \rightarrow inflammation \rightarrow fibrosis\ or\ healing$$

Thus, the process of tissue repair can lead to two different outcomes, depending on the duration and intensity of the injury. In organs with poor ability to regenerate, like the heart, fibrosis is triggered by almost any injury. In organs that can repair themselves to

DOI: 10.1201/9781003356929-10

a certain extent, transient or small injuries lead to healing, but prolonged, repetitive, or extensive injuries cause fibrosis.

Fibrosis has an essential physiological function: if there is a pathogen or a foreign object that cannot be removed, the body tries to encapsulate it in fibrous scar tissue rich with collagen. For example, the hepatitis C virus in the liver causes liver fibrosis (cirrhosis). Likewise, a large wound that cannot be quickly healed needs to be filled in to maintain tissue integrity. Fibrosis does the job.

But fibrosis has a dark side in aging. Tissues progressively tend to show more fibrosis than healing, as we will soon discuss in Part 3 of the book. Fibrosis can cause organs to fail. For example, many types of kidney dysfunction are due to massive scarring of the kidney, and cardiac failure is accompanied by scarring of the heart. Alcoholism leads to liver fibrosis, and liver fibrosis also occurs in non-alcoholic fatty liver disease (NAFLD), associated with obesity, which afflicts ~25% of the world population. A fraction of those with NAFLD progress to chronic inflammation and to fibrosis, with loss of liver function. Fibrosis is also a risk factor for cancer in many organs. There is currently no treatment for fibrosis except for organ transplant. Treating fibrosis is a huge unmet need.

Because inflammation always precedes fibrosis, physicians try to stop inflammation quickly to prevent fibrosis after surgery, stroke, heart attack, and other medical situations. There is usually a **time window** of about 2 days in which stopping inflammation can prevent fibrosis. If the time window is exceeded, fibrosis is inevitable, even if inflammation is stopped. Why this time window? In this chapter, we will try to find out.

Another intriguing question is the slow timescale of healing and scar formation. Despite the brief time window of days we just discussed, it takes *months for the scar to mature* – that is, to reach its final steady-state composition. Likewise, it can take 2 weeks for healing to be completed. Where does this long timescale come from? This is another mystery we will try to explain.

Inflammation and fibrosis is a busy research field in biology and medicine, which has uncovered a large number of molecular facts. Many signaling molecules activate and inhibit immune cells and fibroblasts, and these cells have many possible states.

We will take a big-picture view, putting the essential facts into a mathematical model that captures the core features. This model has a basic property called **multistability** – the ability to produce two or more different stable steady states. Multistability can shed light on how inflammation can lead either to healing, if the injury is brief, or to fibrosis, if the injury is repetitive or prolonged. This understanding also points to potential strategies to prevent and reduce fibrosis.

INFLAMMATION INCLUDES A MASSIVE INFLUX OF IMMUNE CELLS AND ACTIVATION OF MYOFIBROBLASTS

Injury to a tissue causes cells to release factors that cause **inflammation**. Some of the damaged cells die, and others become large metabolically active cells that stop dividing, called senescent cells. The damaged cells and senescent cells secrete "alarm" proteins that flow in the blood like IL6, IL1, and TNF that induce inflammation.

The purpose of inflammation is to fight pathogens and to start repair. Unlike the *T*-cells of the previous chapter, which begin to matter only several days after infection, here we are talking about the innate immune system which is much faster and responds in minutes.

Inflammation has four main features, which are easy to remember by a Latin rhyme – rubor, calor, tumor, dolor: redness, heat, swelling, and pain. The fifth pillar of inflammation is loss of tissue function.

Redness, swelling, and heat are caused by the dilation of nearby blood vessels, which open up to let immune cells flow into the tissue, together with fluids and proteins that fight pathogens (Figure 5.1). The main immune cells are white blood cells that specialize in fighting bacteria called neutrophils. With them come blood monocytes that turn into **macrophages** ("big eaters"), cells that can engulf pathogens and dead cells. They also help to remove the senescent cells, along with other innate immune cells called NK cells. Macrophages play a big role in fibrosis and healing, which we will describe soon.

In parallel to letting in macrophages, injury sets off a process that lays down fibrous material, mainly collagen, to seal up the injury. To do so, damaged cells as well as the incoming macrophages secrete signals including TGF-β that activate a cell type found in every tissue called **fibroblasts** (fiber-forming cells). This signal causes the tissue-resident fibroblasts to proliferate and change their shape to become super-fiber-forming **myofibroblasts** (myo = "muscle-bound"), Figure 5.1. Their muscle-like ability to generate force helps to contract and close the wound.

The two main cell types in our story are thus the incoming macrophages, which we will denote M, and the myofibroblasts, *F*. These two cell types activate each other's proliferation. They do so by secreting **growth factors** for each other – small proteins that diffuse in the tissue and are sensed by receptors on the cell surface. The binding of the growth factor to the receptors makes the cells divide rather than die. Interestingly, *F* cells also secrete a growth factor for themselves, in an example of an **autocrine loop** (Figure 5.2).

This circuit of two cell types was characterized in detail by growing the cells together in a plate by Ruslan Medzhitov and colleagues (Zhou et al. 2018). Growing cells in a plate,

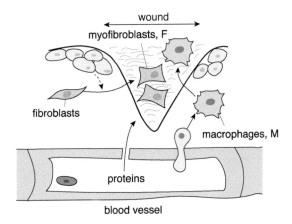

FIGURE 5.1 Inflammation at an injury site involves influx of fluids and macrophages together with differentiation of fibroblasts into myofibroblasts which lay down scar tissue.

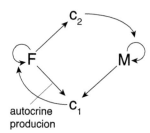

FIGURE 5.2   Macrophages M and myofibroblasts F secrete growth factors c_1 and c_2 for each other, and myofibroblasts also secrete growth factors for themselves in an autocrine loop.

called an "*in vitro*" approach, allows careful measurement of the parameters and dynamics of the circuit, such as cell growth rates and growth-factor secretion and removal rates. Thus, we have estimates for the rate parameters in this circuit.

The situation in the body, *in vivo*, is certainly more complex than in vitro. Still, *in vitro* studies can provide principles to help us understand the *in vivo* process. Generally, in systems medicine there are four approaches: *in vivo*, *in vitro*, *in silico* (computer simulation), and *in envelopo* (back of the envelope calculations, like we do here). Don't expect your friends to know this last term, I invented it for this book.

If the injury is transient, inflammation is resolved within a couple of weeks. M and F cell populations shrink (die by programmed cell death called apoptosis) and vanish. The scar is removed. The tissue cells, such as epithelial cells, divide, and the injury is healed.

If the injury is repetitive or prolonged, however, or if the tissue cannot regenerate (such as heart and brain tissues), M and F populations rise and a permanent scar is formed, made of fibers and cells. This is fibrosis.

Our purpose is to understand the dynamics of the inflammation process and how it can "decide" to show healing or fibrosis.

MATHEMATICAL MODEL FOR MYOFIBROBLASTS SHOWS BISTABILITY

Let's begin by considering only the myofibroblasts, F. This will help us explain the equations and will be useful soon when we add in the macrophages. The main point is that one equation for myofibroblasts can show two different behaviors, a property called bistability.

But first, to clear the mind, let's take a nice deep sigh of relief.

The myofibroblasts produce and secrete a growth factor for themselves, which we will denote c_1 (Figure 5.3). c_1 is degraded at rate γ_1. Thus the rate of change of c_1 is its production minus removal

(5.1)
$$\frac{dc_1}{dt} = a\,F - \gamma_1 c_1$$

The parameters in these equations, based on in vitro data, are as follows: secretion rates like a are about 100 molecules/cell/min, and degradation half-lives $\ln(2)/\gamma_1$ are hours.

Since the production and removal processes take minutes to hours, and cell division and death take a day or longer, we can use **separation of timescales** as in the previous chapters.

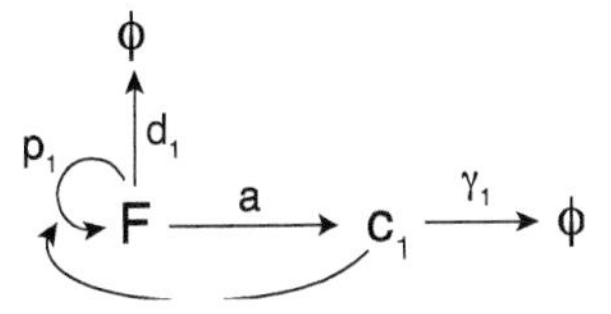

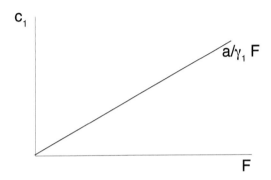

FIGURE 5.3 Circuit describing how myofibroblasts secrete their own growth factor and are removed.

FIGURE 5.4 The concentration of the growth factor is proportional to the concentration of the F cells that produce it at quasi steady state.

On the rapid timescale of hours in which c_1 levels reach their steady state, cell density F hardly changes. Thus, growth factors like c_1 are in quasi-steady-state, which we can compute using $dc_1/dt = 0$. We find that c_1 is proportional to the cells that make it, F (Figure 5.4):

$$(5.2) \qquad c_1 = \frac{a}{\gamma_1} F$$

We now turn to the equation for the rate of change of F cells, given by cell proliferation (all cells come from cells), which increases with c_1, minus cell removal at rate d_1:

$$(5.3) \qquad \frac{dF}{dt} = p_1 F c_1 - d_1 F$$

Plugging in the quasi-steady-state Eq. 5.2 for c_1 results in a proliferation that rises like F^2

$$\frac{dF}{dt} = \frac{p_1 a}{\gamma} F^2 - d_1 F$$

Let's find the fixed points where $dF/dt = 0$. For this purpose, we use the **rate plot**, a useful method for equations with one variable, F in this case. We employed the rate plot for beta cells in Chapter 2. On the x-axis we plot cell density F and on the y-axis we plot the total cell proliferation $\frac{p_1 a}{\gamma} F^2$ that rises quadratically with F and is thus a parabola (black line in Figure 5.5). We next plot the total cell removal $d_1 F$, a line that rises with F (blue line in Figure 5.5).

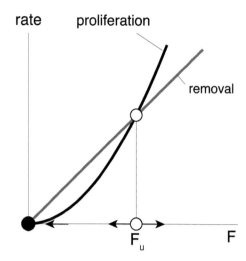

FIGURE 5.5 The total growth rate of F cells rises quadratically since they make their own growth factor. It intersects the removal rate curve that rises linearly with F at two points, a stable fixed point at zero and a second, unstable fixed point.

The interesting points are where the proliferation and death curves cross. These are the **fixed points**. The curves cross at zero and at a higher point, $F_u = \dfrac{d_1 \gamma_1}{p_1 a}$. This point is unstable: F rises to infinity if $F > F_u$, because proliferation exceeds removal (the black line is higher than the blue line). More F cells make more of their own growth factor, spiraling out of control.

Such a rise to infinity is not biologically feasible. We need something to limit F levels. To resolve this, we use the fact that fibroblasts can sense the density of other fibroblasts and stop growing when they are too dense. In a plate, for example, F cells stop dividing when they touch each other. In a tissue they stop growing when they get to a maximal density denoted K. This mechanism prevents fibroblasts from piling up in tissues so as not to gum it up with fibers. The density limit is called a **carrying capacity**. It is an example of law 2, biological processes saturate.

Carrying capacity is modeled in ecology and biology by reducing the proliferation rate when F comes close to carrying capacity K. Proliferation rate is multiplied by the term $(1 - F/K)$. Such a linear reduction term for growth rate, called logistic growth in ecology, is observed experimentally in fibroblasts in vitro (Zhou et al. 2018) by plotting the proliferation rate versus cell number in the plate (blue line in Figure 5.6). Proliferation is measured by a dye that stains cells that replicate their DNA. Indeed, proliferation drops with fibroblast cell density. The intersection of the blue line with the x-axis is the carrying capacity – the cell density that pushes growth rate to zero. Note for future use that macrophages, M, the green data points, seem not to have a measurable carrying capacity.

Thus, our equation with carrying capacity reads

(5.4)
$$\frac{dF}{dt} = \frac{p_1 a}{\gamma} F^2 \left(1 - F/K\right) - d_1 F$$

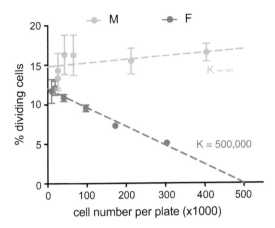

FIGURE 5.6 Cell carrying capacity can be measured by extrapolating the growth rate as a function of cell concentration to determine the concentration of zero growth. Adapted from Zhou et al. (2018).

Let's find the fixed points using a new rate plot (Figure 5.7). On the x-axis we plot cell density F and on the y-axis we plot the total cell proliferation, which now looks like a hill with a dent on the left. The drop of the hill on the right is due to the carrying capacity term that goes to zero when $F = K$. The death curve remains a straight line as before.

If removal rate d_1 is not too large, the removal curve crosses the proliferation curve *three times*: at zero, at a middle concentration, and at a high concentration of cells (Figure 5.7). Let's analyze the three fixed points of Figure 5.7. The middle-fixed point is an **unstable fixed** point, F_u. To see this, note that if F is smaller than F_u, the proliferation curve is lower

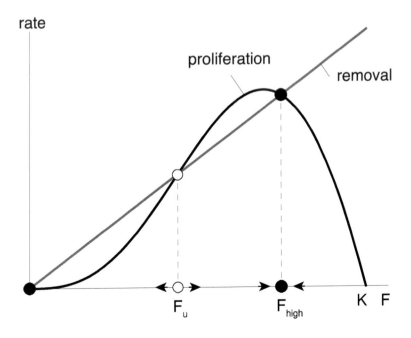

FIGURE 5.7 With carrying capacity, growth and removal curves can cross three times, generating two stable fixed points with an intermediate unstable fixed point.

than death and thus F flows to zero. If F is larger than F_u, F levels flow to the high fixed point, F_{high}. Thus $F = 0$ and $F = F_{high}$ are two **stable fixed points**.

This feature, two stable fixed points for the same equation, is called **bistability**. Depending on initial conditions, the system flows to one of two possible stable states. This can be seen in a plot of F versus time for different initial conditions (Figure 5.8): below an initial level of F_u, F crashes to zero cells; above F_u, F converges to a specific steady-state concentration F_{high} no matter what the starting level was.

Each fixed point has its own **basin of attraction**, defined as the range of initial conditions which flow to that fixed point. Cell density below F_u is in the basin of attraction of the zero fixed point, called the OFF state; above F_u is the basin for the high fixed point, called the ON state.

In vitro, F cells can indeed support themselves at sufficiently high concentrations. The steady state is an ongoing balance of cells dividing and dying about once per day.

Notably, if the death rate is too high, or proliferation is too low, there is only one solution, at zero, as can be seen in the rate plot in (Figure 5.9). The removal and production curves cross only once. A change of parameters causes the loss of a stable fixed point! We will use this fact when we discuss ways to avoid fibrosis.

This loss of bistability occurs when parameters that remove F cells exceed parameters that favor F cells. One can write a ratio of the "pro-F" parameters – the carrying capacity K, autocrine growth factor secretion a and F proliferation rate p_1, relative to factors that are "anti-F" such as death rate d_1 or growth-factor removal rate γ_1. Loss of bistability occurs when the ratio of these pro-F and anti-F parameters goes below a threshold: $p_1 a \, K / \gamma d_1 < 4$ (as shown in Solved Exercise 5.1). Exactly at the threshold, the removal curve touches the proliferation curve at a half-stable point, in addition to the zero fixed point.

THE MACROPHAGE-MYOFIBROBLAST CIRCUIT PROVIDES TWO FIBROSIS STATES AND A HEALING STATE

We now add macrophages, M. The two cell-types together form a circuit that generates bistability, with an OFF state of healing and an ON state of fibrosis. They even have an additional ON/OFF state, which is a second kind of fibrosis.

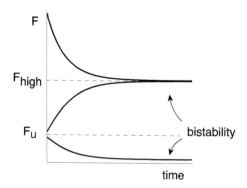

FIGURE 5.8   Bistability means that the same system has two different stable fixed points. Initial cell concentrations determine whether the high or low fixed point is reached.

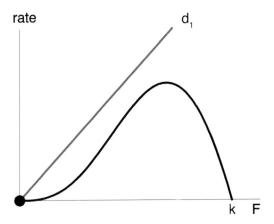

FIGURE 5.9 Bistability is lost suddenly when parameters change and the high stable fixed point vanishes. This occurs when death rate is high or proliferation rate is low.

The macrophages M pour into the tissue from the circulation during inflammation. Note that we ignore details here such as the existence of multiple states of M cells (called $M1$ and $M2$ states, for example), by lumping them together into a single variable $M(t)$. The F cells enhance proliferation of macrophage by secreting an M-specific growth factor. Macrophages M support F proliferation by secreting an F-specific growth factor. This F-specific growth factor described above, so we will group the two growth factors together as c_1. Thus, M and F act to increase each other's numbers (Figure 5.10).

Unlike F cells, the M cells *have a very high carrying capacity*: their numbers can increase by tens of folds when inflammation causes a large influx. They don't approach their very high-carrying capacity in most physiological situations (green line in Figure 5.6). Thus M cells require a different mechanism to avoid spiraling out to very high concentrations. This mechanism is a negative feedback loop due to a basic biological process: *the cells that respond to a growth factor also eat it up.* M cells suck in the receptor on their membrane when it binds to the growth factor. They then degrade the growth factor and sometimes the receptor too – a process called **endocytosis** (Figure 5.11). Endocytosis ensures that if there are too many M cells, they eat up their own growth factor and their numbers thus reduce back to steady-state.[1]

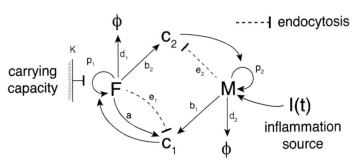

FIGURE 5.10 The macrophage-fibroblast circuit in which both cell types secrete growth factors for each other and deplete the growth factors by endocytosis. Fibroblasts have a carrying capacity and secrete their own growth factors in an autocrine loop.

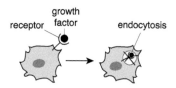

FIGURE 5.11 Endocytosis is the intake of molecules that bind a receptor into the cell resulting in their degradation.

In Solved Exercise 5.2, we show how to derive the fixed points for this circuit, again using separation of timescales and the useful technique of nullclines.

To understand entire dynamics in a single picture we use the phase portrait. The axes are the concentrations of M and F cells. At each point in the plane, we plot a little arrow showing where M and F flow to if they start at that point. The arrows indicate the direction of flow. It's like a snapshot of the dynamics (Figure 5.12).

The phase portrait can be experimentally measured *in vitro*. To do so, Zhou et al. plated cells in many different initial concentrations (initial conditions) in a 96-well plate. They watched how the cell concentrations changed over 2 days (Zhou et al. 2018), plotting little arrows to indicate the changes. This provided a phase portrait in one fell swoop!

The phase portrait (see Solved Exercise 5.2) reveals two stable fixed points. There is a fixed point at zero cells ($F=0$, $M=0$). This is the OFF state. It corresponds to healing, since the myofibroblasts and macrophages are gone.

The other fixed point has high levels of both M and F cells, which sustain each other. We call it the ON state (Figure 5.12). The ON state is a stable fixed point. All arrows in the vicinity flow to it. If a perturbation around the ON state occurs, say that a few extra M arrive, they eat up their own growth factor and cell numbers drop back to steady state.

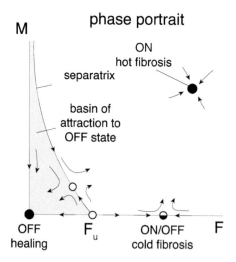

FIGURE 5.12 The phase portrait shows a stable ON state with both cell types, called hot fibrosis, a stable OFF state with neither, and a semi-stable state with only fibroblasts called cold fibrosis. A separatrix divides the phase portrait into basins of attraction for the two stable states, the ON and OFF states.

FIGURE 5.13   Scar fibers *E* are extracellular matrix proteins secreted by fibroblasts, *F*. Macrophages *M* secrete proteins that degrade and remodel the scar.

This is how a molecular process, endocytosis, can provide a systems-level effect of stabilizing fixed points. The carrying capacity for *F* is also essential to stabilize the ON state – without it, both cells would rise indefinitely.

The general condition for stability of such two-cell circuits was defined by Miri Adler et al. (2018). Either (1) both cell types have a carrying capacity or (2) one cell type has a carrying capacity and the other has negative feedback on its growth factor through a mechanism such as endocytosis. The latter applies to the current situation.

To understand fibrosis, we further need to consider the fibers, namely the **extracellular matrix (ECM or E for short)** deposited by *F* cells. In contrast, *M* cells produce molecular scissors (matrix metalloproteinases, or MMPs) that cut *E* up (Figure 5.13). These scissors are also produced at small amounts by the regular cells of the tissue. Thus, *E* rises with *F* and drops with *M*.

The OFF state, in which $F = M = 0$, is the healing state. The fibers *E* go to zero. The ON state, in which both cell types are at a high steady-state concentration, corresponds to fibrosis. The fibers *E* reach a high steady-state concentration, continually made by *F* and degraded by *M*. The fibrotic scar is a living tissue.

The phase portrait shows another interesting fixed point. This is the ON/OFF state with only fibroblasts. This state can be called "cold fibrosis" where cold means no immune cells, $M=0$. The ON state can be called "hot fibrosis." These are new terms for pathology, borrowed from cancer biology in which hot tumors have more immune cells than cold tumors.

The cold fibrosis fixed point can be stable for certain ranges of the circuit parameters, such as weak secretion by *F* cells of growth factors for *M* cells (see Exercise 5.6). One can expect that cold fibrosis is "worse" because there is a lack of *M* cells with their molecular scissors. The ECM is more abundant and stiff than in the ON state. Such stiff and abundant ECM in cold fibrosis is found in end-stage liver cirrhosis called "burnout NASH" and in fibrosis after heart attacks.

Examples of both hot and cold fibrosis states can also be found in the skin. Dermatology recognizes two main types of scars: keloid scars with abundant macrophages (hot fibrosis), and hyperproliferative scars which eventually lose most of their macrophages (cold fibrosis).

INJURY AND INFLAMMATION CAN BE MODELED BY A TRANSIENT INFLUX OF MACROPHAGES

To see the dynamics of healing, let's consider an injury at $t=0$. There is a small initial number of *F* and *M* cells at the injury site. Inflammation can be modeled as a large influx *I(t)* of *M* cells, where *I* stands for "influx."

Consider a 2-day pulse of inflammation in which influx $I(t)$ is high for 2 days and then returns to zero. We again use the phase portrait, but this time in log scale so that we can more easily see the region of low cell concentrations. M levels rise sharply and produce c_1. As a result, F cells begin to divide.

If the dynamics stay within the basin of attraction to the OFF states, M levels fall, and with them F levels, until $F=0$, $M=0$ is reached (Figure 5.14). This trajectory is typical of proper healing. Scar fibers E are deposited by F cells, and when the F cells are gone, scar is degraded by the tissue. Scar $E(t)$ rises and then vanishes (Figure 5.14). The timescale, using the typical parameters of Table 5.1, is about 2 weeks.

Now consider a longer pulse of inflammation which lasts for 4 days. M levels rise sharply and cross the boundary to the basin of attraction to the ON state (Figure 5.15). This boundary is called the **separatrix**. Now there are enough F and M cells to support each other. The cells flow to the ON state. They create a scar tissue with constant turnover of M and F cells, and a high steady-state level of fibers E. Thus, a 4-day inflammation event leads to fibrosis.

Similarly, consider a repeated injury. A 2-day inflammation pulse is not sufficient to cross the separatrix, but if another 2-day pulse occurs after a week, there are enough M and F cells left from the first injury to cross the separatrix and go to fibrosis (Figure 5.16).

Thus, the same system can result in either healing or fibrosis, depending on the strength and duration of the inflammation pulse. The system has a healing state with zero M and F cells and no scar. It has a fibrosis state with lots of F and M cells and permanent scars.

THE TIME WINDOW FOR STOPPING INFLAMMATION IS DUE TO BISTABILITY

We can now understand why it is so urgent to stop inflammation to avoid fibrosis. Let's plot how the duration of the inflammation pulse affects the final (steady state) amount of scar fibers E (Figure 5.17). We see that below a critical duration of inflammation, of about

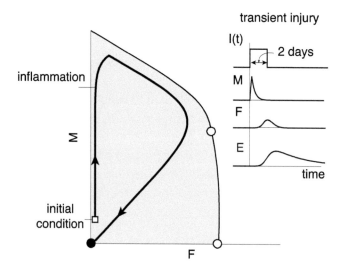

FIGURE 5.14 A transient injury causes an influx of macrophages and rise of fibroblasts but remains in the basin of attraction of the OFF state so that cell populations decline and the scar is removed.

TABLE 5.1 Representative Parameter Values for Cell Circuits

Parameter	Biological Meaning	Value
p_1	Maximal proliferation rate of myofibroblasts	0.9 day^{-1}
p_2	Maximal proliferation rate of macrophages	0.8 day^{-1}
d_i	Removal rate of the cells	0.3 day^{-1}
K	Carrying capacity of myofibroblasts	10^6 cells ($\sim 10^{-3} \frac{\text{cell}}{\mu\text{m}^3}$)
k_i	Binding affinity of growth factor c_i	6×10^8 molecules
b_2	Maximal secretion rate of growth factor by myofibroblasts	$470 \frac{\text{molecules}}{\text{cell min}}$
b_1	Maximal secretion rate of growth factor by macrophages	$70 \frac{\text{molecules}}{\text{cell min}}$
a	Maximal autocrine secretion rate by myofibroblasts	$240 \frac{\text{molecules}}{\text{cell min}}$
e_2	Maximal endocytosis rate of growth rate by macrophages	$940 \frac{\text{molecules}}{\text{cell min}}$
e_1	Maximal endocytosis rate of growth rate by myofibroblasts	$510 \frac{\text{molecules}}{\text{cell min}}$
γ	Degradation rate of growth factors	2 day^{-1}

Source: Adapted from Adler et al. (2018).
Note: This table uses the more accurate parameters for Michaelis–Menten functions for endocytosis and proliferation, which go as $c_i/(k_i + c_i)$ instead of as linearly with c_i as in the text.

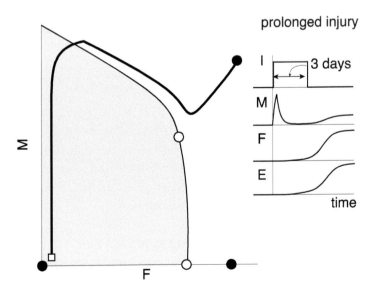

FIGURE 5.15 A prolonged injury crosses the separatrix into the basin of attraction of the ON state, causing fibrosis in which fibroblasts and macrophages support each other and continuously turn over.

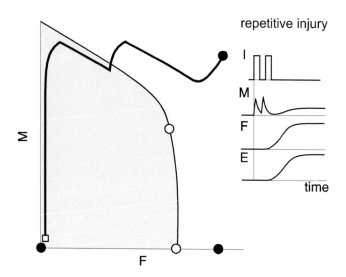

FIGURE 5.16   Recurring or repetitive injury can cross the separatrix and trigger by fibrosis.

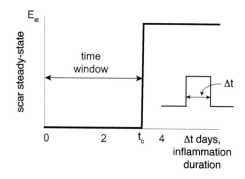

FIGURE 5.17   A time window for preventing fibrosis: inflammation lasting less than about 3 days avoids fibrosis whereas fibrosis is reached when inflammation lasts longer than 3 days.

$t_c = 3$ days, the scar vanishes. Above t_c, M crosses the separatrix and the ON state is inevitable *even if the inflammation stops.*

For example, an inflammation pulse that lasts a bit longer than the critical threshold t_c causes M to cross the separatrix, and F-cells have a bit of time to multiply. When inflammation stops, F cells are not very many and hence M levels sharply drop (Figure 5.16). But just before they crash to zero, they recover due to the increased F cells that are just enough to support them. Both F and M go up together over weeks to the ON state. Fibrosis occurs even though the inflammatory pulse stopped long ago.

The main circuit property here, bistability, is worth remembering. It can potentially explain other medical situations in which there is a limited time window to prevent irreversible outcomes. Two deadly examples are septic shock and hemorrhagic shock, in which a bacterial infection or hemorrhage causes lowered blood pressure, blood clotting, and organ failure. There is a golden hour to treat shock and sepsis, with a turning point: some patients slowly recover whereas others quickly plummet.

THE LONG TIMESCALE FOR SCAR MATURATION AND HEALING IS DUE TO THE SLOWDOWN NEAR AN UNSTABLE FIXED POINT

Scar maturation is a process that unfolds over months; the scar changes until it reaches steady state. This timescale is much slower than the cell turnover time of days. How does the long timescale arise?

The slow timescale found in the model is due to the fact that the dynamics near the separatrix approach an unstable fixed point. This is the white circle in the middle of the separatrix in Figures 5.14–5.16. By definition, at a fixed point, including an unstable one, the velocity is zero (no change). Thus, the velocity is always slow near a fixed point, causing a slowdown phenomenon.

Intuitively, the slowdown is similar to a ball trying to climb out of a valley and go over a ridge. The ball slows as it approaches the summit and then speeds up again (Figure 5.18). The summit is an unstable fixed point, any perturbation makes the ball roll away. The same applies to the healing process, which dawdles around the unstable fixed point and takes about 2 weeks to resolve back to the OFF state.

STRATEGIES FOR PREVENTING AND REVERSING FIBROSIS

A general cure for fibrosis has not yet been achieved, and many attempts have failed. So, let's use what we have learned to explore what future interventions might prevent fibrosis.

To prevent fibrosis, we need to enlarge the basin of attraction for the OFF (healing) state. A large basin means that more situations end up resolved without fibrosis (Figure 5.19). We can also explore whether fibrosis can be reversed: can a mature scar in the ON state be made to flow to the OFF state.

Evidence for reversal of long-standing fibrosis has been accumulating, revolutionizing thinking in the field. Fibrosis can vanish in the liver, for example, after successful antiviral treatment of hepatitis C infection. Such reversal of fibrosis depends on the fact that fibrosis is a dynamic steady state with cell turnover. Similarly, fibrosis seems to be tunable in different biological contexts. Embryos do not show fibrosis after injury, and many mammals regenerate more readily than humans, giving hope that tweaking the circuit can help prevent or abrogate fibrosis.

We can use our circuit model to scan for parameter changes that affect fibrosis. Miri Adler did so and found that prevention and reversal of fibrosis become possible in the

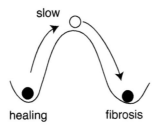

FIGURE 5.18 Dynamics are slow near an unstable fixed point, in analogy to a ball rolling to the top of a hill.

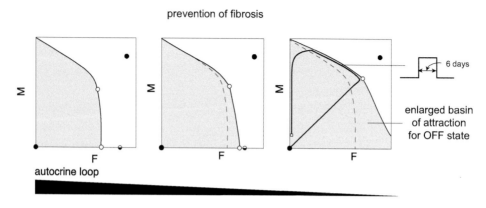

FIGURE 5.19 Strategy to resolve fibrosis by inhibiting the autocrine loop of fibroblasts. This enlarges the basin of attraction to the healing state.

model when the ON/OFF fixed point vanishes, because this greatly expands the basin of attraction of the OFF state. In other words, one must eliminate the fixed point in which F cells support themselves. To do so, as we saw in Figure 5.11, requires a combination of parameters to go below a threshold, the ratio of pro-F to anti-F parameters:

$$p_1a \; K/\gamma d_1 < 4$$

This parameter group offers several targets against fibrosis. The equation gives hope that one does not need to push a parameter all the way down to zero, which is difficult. Instead, one must merely nudge the system below a threshold in order to collapse fibrosis.

You can see what happens in Figure 5.19 when you reduce this parameter group – the unstable fixed point (white circle on the x axis) moves to higher and higher values, enlarging the basin of attraction to the OFF state. At a certain value, the unstable fixed point collides with the ON/OFF fixed point, and both fixed points annihilate! The basin of attraction is now very large.

The dynamics when this anti-fibrosis condition is met is exemplified in Figure 5.19. A lengthy inflammation pulse of 6 days which would normally lead to fibrosis now flows to the OFF state with no fibrosis. Longer pulses of 8 days can still result in ON-state fibrosis.

One target suggested by the circuit equation is to inhibit the growth factor for F cells which the F cells themselves secrete – the autocrine growth factor for myofibroblasts. This can be achieved by increasing its removal rate γ or reducing its production rate a; both changes push the parameter group down, as required.

Shoval Miyara and Eldad Tzahor took on this challenge, using a well-established model for fibrosis, namely heart attacks. Shoval identified a major autocrine factor of mouse heart myofibroblasts, a growth factor called TIMP1. This growth factor is not expressed in the normal uninjured heart. He injected antibodies against this autocrine factor, thereby inactivating it. Mice showed much less scarring in experimentally induced heart attacks (Miyara 2023).

Similarly, this theory inspired Shuang Wang in Scott Friedman's lab to inhibit the liver myofibroblast autocrine loop. She achieved a reduction of advanced liver fibrosis in mice, by inhibiting an autocrine receptor for liver myofibroblasts, NTRK3 (Wang, 2023).

Thus, this circuit-to-target approach may help to address fibrosis in different organs - heart attacks which are an acute injury, and NASH-induced liver fibrosis which is a chronic condition.

Here we see several benefits of math modeling. It can give hope and guidance. We do not need to kill all the fibroblasts. We just need to push certain rates enough so that they go below a threshold. Another benefit of a good theory is that it can inspire new experiments, as in the experiments on heart and liver. Theory provides new concepts – in this case, the concepts of hot and cold fibrosis that provide new ways to analyze tissue samples. Their utility was recently demonstrated in the kidney in which hot and cold fibrosis regions can coexist depending on oxygen and inflammation levels (Setten et al. 2022). This theory also helped to start experiments on the cancer micro-environment, the support system for cancer composed of F and M cells, aiming to collapse the support in order to fight cancer. Can't wait to see the results.

EXERCISES

Solved Exercise 5.1: Find the condition for bistability in the model for fibroblasts, Eq. 5.4.

Solution: Our equation is $\dfrac{dF}{dt} = \dfrac{p_1 a}{\gamma} F^2 (1 - F/K) - d_1 F$

The fixed points occur at $F=0$ and at two, one, or zero other points determined by whether the removal line $d_1 F$ intersects the proliferation "hill." To solve for those non-zero fixed points, we can set $dF/dt = 0$ and divide by F (since we assume F is non-zero) to find

$$d_1 = \frac{p_1\,a}{\gamma} F\left(1 - \frac{F}{K}\right).$$

To simplify things, let's divide and multiply by K, and divide by d_1 so that

$$1 = \left(\frac{p_1\,a\,K}{\gamma\,d_1}\right) \frac{F}{K}\left(1 - \frac{F}{K}\right).$$

We did that because the term $F/K(1 - F/K)$ is easy: it's a symmetric parabola that is zero at $F = 0$ and $F = K$. Its maximum value occurs between the two roots at $F = K/2$, where its height is 1/4 (Figure 5.20). Thus, the condition for two non-zero fixed points is

$$\frac{p_1\,a\,K}{\gamma\,d_1} > 4.$$

A single "half-stable" fixed point occurs when this precisely equals 4 (try to analyze this case).

Solved Exercise 5.2: Find the fixed points of the two-cell circuit composed of fibroblasts and macrophages.

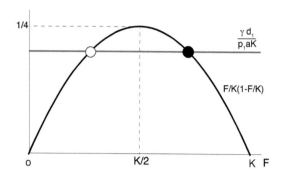

FIGURE 5.20 Rate plot for the growth and removal rates of fibroblasts shows how the number of fixed points depends on parameters.

Solution: Let's write the equations for this circuit. The F-specific growth factor c_1 is secreted by both M and F cells, and endocytosed by its receivers, F cells:

$$\frac{dc_1}{dt} = aF + b_1 M - e_1 F c_1 - \gamma_1 c_1$$

where b_1 is production rate per M cell. The M-specific growth factor, c_2, is produced by F and endocytosed by M cells:

$$\frac{dc_2}{dt} = b_2 F - e_2 M c_2 - \gamma_2 c_2$$

endocytosis rates like e_1, e_2 are about 1000/cell/min. Therefore, endocytosis is the main removal mechanism of growth factors unless cell density is very low. Growth factors dynamics have timescales of minutes to hours, whereas cell growth is much slower with a timescale of days. We thus invoke separation of timescales and compute the quasi-steady-state of the two growth factor concentrations:

$$c_1 = \frac{aF + b_1 M}{e_1 F + \gamma_1}, \quad c_2 = \frac{b_2 F}{e_2 M + \gamma_2}$$

The M cells divide under control of c_2. Unlike F cells, the M cells *are far from their carrying capacity*. Thus, M cells follow the simple equation

$$\frac{dM}{dt} = p_2 M c_2 - d_2 M$$

The F-cell equation is as above, Eq. 5.4. Plugging in the quasi-steady-state values for c_1 and c_2, we arrive at the cell equations on the scale of cell turnover (days)

$$\frac{dM}{dt} = M \left(p_2 \frac{b_2 F}{e_2 M + \gamma_2} - d_2 \right)$$

$$\frac{dF}{dt} = F\left(p_1 \frac{a\,F + b_1 M}{e_1 F + \gamma_1}\left(1 - \frac{F}{K}\right) - d_1 \right)$$

Looks a bit complicated… but we can make progress. To understand these equations, we use the method of **nullclines** – the lovely graphical method we've seen in previous chapters. Nullclines are the extension of the rate plot approach. Whereas rate plots work well for a single variable, nullcline are helpful for systems of equations with two variables such as $M(t)$ and $F(t)$.

Nullclines are curves in which one of the two cell concentrations does not change. One nullcline is $dM/dt = 0$, and the other is $dF/dt = 0$. The fixed points are where the two nullclines intersect, because at fixed points both cell populations don't change. It's therefore useful to draw both nullclines on the phase plane, whose axes are F and M cell concentrations, and study the intersection points.

The $dM/dt = 0$ nullcline is composed of the x-axis, $M = 0$, and of the solution to $p_2 \dfrac{b_2 F}{e_2 M + \gamma_2} - d_2 = 0$. The latter is a straight line, $M = \alpha\,F - \beta$, with an intercept $\beta = \dfrac{\gamma_2}{e_2}$. The intercept is close to zero because endocytosis dominates degradation and thus $\beta \ll 1$. Plotting this line separates the phase plane into two regions, a top region in which M drops and a bottom region in which M rises (Figure 5.21).

The $dF/dt = 0$ nullcline is the y-axis $F = 0$ and the solution to

$$p_1 \frac{aF + b_1 M}{e_1 F + \gamma_1}\left(1 - F/K\right) - d_1 = 0.$$

The $F = 0$ and $M = 0$ nullclines intersect at zero, which is the OFF state. Zero cells is a stable state since at very low cell numbers there is not enough c_1 and c_2 to overcome cell removal, and both cell populations crash.

The more complicated F-nullcline equation can be understood if we look at the $M = 0$ line. There, we have the three fixed points we saw in exercise 1 when we discussed F alone. Plotting the nullcline, which looks like $M \sim (F + \gamma_1/e_1)/(1 - F/K) - a/b_1 F$, we see that it has a U-shape which drops through the unstable fixed point F_u, drops below zero and rises

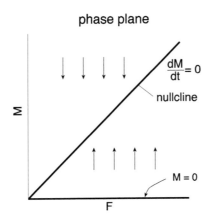

FIGURE 5.21 Each nullcline $dX/dt = 0$ divides the phase plane into regions where the relevant variable X moves in one direction.

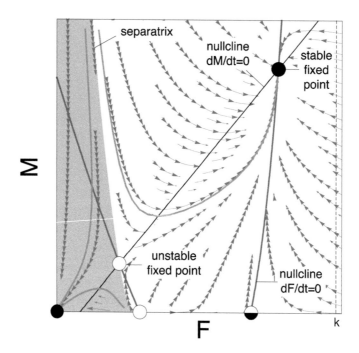

FIGURE 5.22   Nullclines and phase portrait for the fibroblast-macrophage circuit. Selected trajectories are shown in orange.

through the high fixed point F_{high}, and then climbs up and diverge near the carrying capacity $F = K$ (blue curve in Figure 5.22). The orange curves on the phase portrait show three different trajectories.

The phase portrait indicates that the zero and high stable fixed points are stable (arrows flow into them). An analytical method called linear stability analysis can be used to confirm which points are stable (black dots) and which are unstable (white dots), and which are half stable (half-white half-black dots).

Exercise 5.1: Nullclines and directions of motion

The nullcline $dM/dt = 0$ is the line where M does not change. On one side of the nullcline in phase plane, $dM/dt > 0$ which means that M grows, and on the other side, $dM/dt < 0$ which means that M shrinks

a. Why is this statement true?

b. Which side of the nullcline corresponds to $dM/dt > 0$ and which to $dM/dt < 0$?

c. Repeat for the $dF/dt = 0$ nullcline. Explain why this U-shaped nullcline separates the phase plane to a middle region where F flows to higher levels, and regions at low and high F where F flows to lower levels.

d. Use these results to sketch the arrows in the phase portrait and to explain the stability of the fixed points.

Exercise 5.2: Saturating growth factors

Repeat the calculation when c_1 and c_2 act on F and M in a Michaelis–Menten way $c_1 / (k_1 + c_1), c_2 / (k_2 + c_2)$. The same terms appear in the endocytosis term because binding of growth factor to its receptor both initiates the signaling that affects proliferation and leads to endocytosis.

Exercise 5.3: Paradoxical effects of macrophage depletion

Experiments that deplete macrophages show conflicting effects on fibrosis. In some contexts, fibrosis is prevented whereas in others it is enhanced (Duffield et al. 2013). Show, using the model of Solved Exercise 5.2, that the deciding factor is the timing of depletion.

  a. What happens when M is set to zero with fibroblasts below their unstable fixed point, $F < Fu$?

  b. What happens when M is set to zero with fibroblasts below their unstable fixed point, $F > Fu$?

Exercise 5.4: Extracellular matrix (ECM) accumulation in tissue repair and fibrosis

ECM is produced by myofibroblasts. ECM degradation is controlled by proteins called MMPs and TIMPs, where MMPs enhance the degradation of ECM and TIMPs inhibit the degradation of ECM. MMPs are produced mainly by macrophages apart from a small baseline level that is produced by the tissue. TIMPs are produced by both macrophages and myofibroblasts (Figure 5.23).

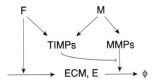

FIGURE 5.23   Circuit in which fibroblasts F produce extracellular matrix E, which is degraded by MMP proteins secreted by macrophages M. These proteins are inhibited by TIMP proteins secreted by fibroblasts.

a. Follow the interactions above to explain each term in the following set of differential equations for MMPs, TIMPs, and ECM.

$$\frac{dMMP}{dt} = \epsilon + a\ M - \alpha_1 MMP$$

$$\frac{dTIMP}{dt} = b\ M + c\ F - \alpha_2 TIMP$$

$$\frac{dECM}{dt} = d\ F - \alpha_3 \frac{MMP}{TIMP + k} ECM$$

b. Assuming that the factors that control ECM degradation reach steady state faster than ECM, rewrite the equation for ECM with the steady states of MMPs and TIMPs.

c. Solve the steady state of ECM and describe its dependence on the number of myofibroblasts and macrophages.

d. What are ECM steady states in healing, hot fibrosis, and cold fibrosis (don't solve the cell steady states, just use steady-states notation such as F-hot for myofibroblasts level in hot fibrosis)? What can you say about the dependence of the scar size on F in hot fibrosis versus cold fibrosis if you consider that myofibroblasts numbers are approximately the same in the two fibrotic states?

Exercise 5.5: Diffusion range of growth factors due to endocytosis (Oyler-Yaniv et al. 2017)

A growth factor c diffuses with a diffusion coefficient D and is endocytosed (removed) by cells with density F at a rate $e\ F\ c$.

a. How long can the molecule travel on average before being removed? Show that this is approximately $L = \sqrt{\frac{D}{eF}}$. Show that this length scale is about 100 microns for typical diffusion constants and cell densities.

b. Suppose the density of target cells F is low. How does this affect the range? What are the consequences for biological regulation of cell circuits?

c. Suppose that two micro-injuries of diameter 50 micron are made in a tissue at a distance of r from each other. Intuitively guess how the response would differ if r is much larger than L or similar to L?

Exercise 5.6: Stability of cold fibrosis

Analyze the fibrosis circuit of Solved Exercise 5.2, and weaken the secretion rate of secretion of the macrophage growth by myofibroblasts, described by the parameter b_2.

a. At which value of b_2 does the cold fibrosis fixed point become stable?

b. What happens to the hot fibrosis fixed point? Sketch the nullclines and the fixed points.

c. What happens if $b_2 = 0$?

d. What are the implications for considering the macrophage growth factor, or macrophages themselves, as a target for an anti-fibrosis treatment?

NOTE

1 Endocytosis also provides a length scale of about 10–100 μm, or about one to ten cell diameters, for the distance a secreted molecule diffuses before it is eaten up by its target cells (Oyler-Yaniv et al. 2017). This provides a natural compartment size for cell-cell circuits. The lower the target cell density, the longer this range, because there is less endocytosis, ensuring that the secreted molecule reaches its target (see Exercise 5.5).

REFERENCES

Adler, Miri, Avi Mayo, Xu Zhou, Ruth A. Franklin, Jeremy B. Jacox, Ruslan Medzhitov, and Uri Alon. 2018. "Endocytosis as a Stabilizing Mechanism for Tissue Homeostasis." *Proceedings of the National Academy of Sciences* 115 (8): E1926–35. https://doi.org/10.1073/pnas.1714377115.

Duffield, Jeremy S., Mark Lupher, Victor J. Thannickal, and Thomas A. Wynn. 2013. "Host Responses in Tissue Repair and Fibrosis." *Annual Review of Pathology* 8 (January): 241–76. https://doi.org/10.1146/annurev-pathol-020712-163930.

Oyler-Yaniv, Alon, Jennifer Oyler-Yaniv, Benjamin M. Whitlock, Zhiduo Liu, Ronald N. Germain, Morgan Huse, Grégoire Altan-Bonnet, and Oleg Krichevsky. 2017. "A Tunable Diffusion-Consumption Mechanism of Cytokine Propagation Enables Plasticity in Cell-to-Cell Communication in the Immune System." *Immunity* 46 (4): 609–20. https://doi.org/10.1016/j.immuni.2017.03.011.

Setten, E., A. Castagna, J.M. Nava-Sedeño, et al. 2022. "Understanding Fibrosis Pathogenesis Via Modeling Macrophage-Fibroblast Interplay in Immune-Metabolic Context." *Nature Communications* 13: 6499. https://doi-org.ezproxy.weizmann.ac.il/10.1038/s41467-022-34241-5

Zhou, Xu, Ruth A. Franklin, Miri Adler, Jeremy B. Jacox, Will Bailis, Justin A. Shyer, Richard A. Flavell, Avi Mayo, Uri Alon, and Ruslan Medzhitov. 2018. "Circuit Design Features of a Stable Two-Cell System." *Cell* 172 (4): 744–57.E17. https://doi.org/10.1016/j.cell.2018.01.015.

Miri Adler, Avi Mayo, Xu Zhou, Ruth A Franklin, Matthew L Meizlish, Ruslan Medzhitov, Stefan M Kallenberger, Uri Alon "Principles of Cell Circuits for Tissue Repair and Fibrosis". iScience. 2020 Feb 21;23(2):100841. doi: 10.1016/j.isci.2020.100841. Epub 2020 Jan 16. PMID: 32058955; PMCID: PMC7005469.

Altan-Bonnet, Gregoire, and Ratnadeep Mukherjee. 2019. "Cytokine-Mediated Communication: A Quantitative Appraisal of Immune Complexity." *Nature Reviews Immunology* 19 (4): 205–17.

Arno, A. I., G. G. Gauglitz, J. P. Barret, and M. G. Jeschke. 2014. "Up-to-Date Approach to Manage Keloids and Hypertrophic Scars: A Useful Guide." *Burns* 40 (7): 1255–66. https://doi.org/10.1016/j.burns.2014.02.011.

Love, P. B., and R. V. Kundu. 2013. "Keloids: An Update on Medical and Surgical Treatments." *Journal of Drugs in Dermatology* 12 (4): 403–9.

Medzhitov, R. 2021. "The Spectrum of Inflammatory Responses." *Science* 374: 1070–5. https://www.science.org/doi/full/10.1126/science.abi5200.

Miyara, Shoval, Miri Adler, Elad Bassat, Yalin Divinsky, Kfir B. Umansky, Jacob Elkahal, Alexander Genzelinakh, David Kain, Daria Lendengolts, Tali Shalit, Michael Gershovits, Avraham Shakked, Lingling Zhang, Jingkui Wang, Danielle M. Kimchi, Andrea Baehr, Rachel Sarig, Christian Kupatt, Elly M. Tanaka, Ruslan Medzhitov, Avi Mayo, Uri Alon, Eldad Tzahor. 2023. "Circuit to Target Approach Defines an Autocrine Myofibroblast Loop That Drives Cardiac Fibrosis." *bioRxiv*. doi: https://doi.org/10.1101/2023.01.01.522422

Wang, S., K. Li, E. Pickholz, R. Dobie, K.P. Matchett, N.C. Henderson, C. Carrico, I. Driver, M. Borch Jensen, L. Chen, M. Petitjean. 2023. "An Autocrine Signaling Circuit in Hepatic Stellate Cells Underlies Advanced Fibrosis in Nonalcoholic Steatohepatitis." *Science Translational Medicine* 15(677): eadd3949.

Zhou, Y., H. Zhang, Y. Yao, X. Zhang, Y. Guan, and F. Zheng. 2022. "CD4+ T Cell Activation and Inflammation in NASH-Related Fibrosis." *Frontiers in Immunology* 13: 967410.

III

Aging and Age-Related Diseases

CHAPTER 6

Basic Facts of Aging

WELCOME TO PART 3! This part is devoted to the fascinating topic of aging. We will use our three laws to develop a theory of aging and test it against a wide range of experiments.

In this chapter, I curated many quantitative patterns of aging. Such patterns are the basis for forming and testing theoretical understanding. In the coming chapters, we will use these patterns to develop fundamental principles for the causes and rates of aging and the origins of aging-related diseases.

AGING IS DEFINED BY RISK OF DEATH AND DISEASES THAT RISES WITH AGE

To understand aging, and to introduce some of the basic concepts, let's begin with hypothetical organisms that do not age. Consider a group of these organisms that are killed by predators at a constant rate, h_0, regardless of age. The parameter h_0 is called **extrinsic mortality**. Over time, there remain fewer and fewer organisms,

$$\frac{dN}{dt} = -h_0 N.$$

The solution is exponential decay,

$$N(t) = N(0)e^{-h_0 t},$$

where $N(0)$ is the initial population.

The **survival curve** for this population, defined as the fraction of organisms that remain alive at time t, thus decays exponentially

$$S(t) = N(t)/N(0) = e^{-h_0 t}.$$

This is just like radioactive decay of particles (Figure 6.1). The probability of death per unit time, called the **hazard**, is independent on age, $h(\tau) = h_0$ (Figure 6.2). This is what "no aging" looks like, in terms of survival curves.

DOI: 10.1201/9781003356929-12

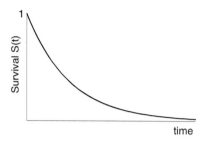

FIGURE 6.1   The survival curve – the fraction surviving to a given age – for organisms that do not age and have constant extrinsic mortality.

FIGURE 6.2   The hazard curve – probability of death per unit time – for the organisms of Figure 6.1 is independent on age.

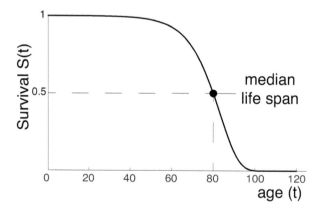

FIGURE 6.3   The human survival curve shows a death rate that increases with age.

Let's now consider the human survival curve (Figure 6.3). It does not decay exponentially. Instead, death is delayed on average: the survival curve starts out nearly flat. Death is rare until the seventh decade, and then death becomes common. Risk of death is age dependent.

AGING HAS NEARLY UNIVERSAL FEATURES

The hazard curve allows us to see more details. It is defined as the fraction of individuals that die at a given age out of all those that survive to that age. Or mathematically, $h = -\dfrac{1}{S}\dfrac{dS}{dt}$. The human hazard curve has an interesting shape (Figure 6.4).

This data is for Sweden in 2012, and similar graphs are found across the world. Risk of death is high in the first year: the human life cycle begins with rapid growth of the embryo with accompanying diseases and delivery risks. Some infant diseases arise from mutations in the germline which cause rare congenital disorders; over 6000 known genetic disorders together account for a mortality rate on the order of 10^{-3} in the first year.

The hazard curve drops to a minimum during childhood. In the teenage years, hazard rises again, and plateaus in early adulthood. In this plateau, hazard is dominated by extrinsic mortality: accidents, suicides, and homicides at a rate of about 3 out of 10,000 per year. Then, starting at age 30, risk of death begins to rise sharply and doubles about every 8 years. This exponential rise in hazard is called the **Gompertz law**. If we denote age by τ, the Gompertz law is

$$h(\tau) \sim be^{\alpha\tau}.$$

where α is the slope of log incidence called the Gompertz slope and b is the intercept. This law was discovered by Benjamin Gompertz in 1825, a mathematician who found work computing life-expectancy tables for an insurance company. If we separate mortality into intrinsic and extrinsic components, we can see that the exponential rise in intrinsic hazard begins already around age 15, as in Figure 6.5 (Carnes et al. 2006) that shows US mortality statistics for males and females.

Different regions and historical periods differ mainly in their extrinsic mortality and childhood mortality. In past centuries, and in some countries today, childhood mortality is about 20% and about 1% of mothers die at childbirth. The Gompertz slope α is,

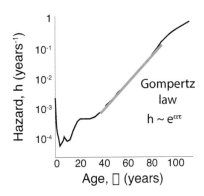

FIGURE 6.4 Human hazard curve shows an exponentially rising probability of death with age known as the Gompertz law, with deceleration at very old ages. Adapted from Barbieri et al. (2015).

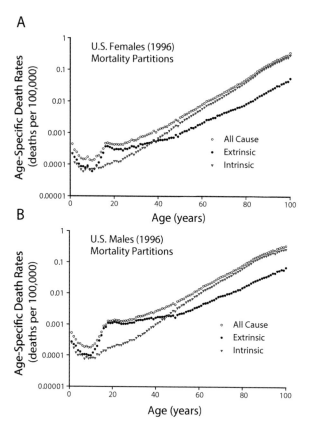

FIGURE 6.5 Hazard curves show that intrinsic mortality rises exponentially from teenage and on, and exceeds extrinsic mortality in adulthood and old age Adapted from Carnes et al. (2006).

however, more constant across populations. Thus, hazard curves are often modeled by the Gompertz–Makeham law that adds extrinsic mortality h_0

$$h(\tau) \sim be^{\alpha\tau} + h_0.$$

Extrinsic mortality rises with age as seen in Figure 6.5, perhaps because accidents become more lethal. It rises more slowly than intrinsic mortality.

The Gompertz law is nearly universal. It is found in most animals studied. This includes the favorite model organisms of laboratory research: mice that live for about 2.5 years, *Drosophila* fruit flies that live for about 2 months, and *C. elegans* roundworms that live about 2 weeks. In 2019, Yifan Yang et al. (2019) found that the Gompertz law holds even in *E. coli* bacteria: when starved, their risk of death, measured by a dye that enters dead cells, rises exponentially with an average lifespan of about 100 hours.

There are exceptions to the rule, such as some trees in which hazard drops with age. Some organisms have very low intrinsic mortality, such as hydra that grows indefinitely, or cells that divide symmetrically such as bacteria in rich medium.

Another nearly universal feature is that the exponential Gompertz law slows down at very old ages, around age 80 in humans. The hazard curve begins to flatten out. Above age 100 hazard is believed to plateau at about a 50% chance of death per year.

Thus, aging means that there is something different about young and old organisms. Age 20 and 70 is different. Something accumulates or changes in the body to make the hazard curve rise sharply with age.

Indeed, most physiological and cognitive functions decline with age. This includes physical ability and organ function (Figure 6.6), male and female reproductive capacity, as well as vision, hearing and aspects of cognitive ability (Figure 6.7). It is worth noting that organs have spare capacity: you can remove 90% of the pancreas or kidneys and survive (although you lose resilience to stress). That is why people can donate a kidney and remain healthy. Because organs compensate for damage before they begin to lose function, the pathological consequences of the decline are felt primarily at old age when spare capacity is used up.

Not everything declines; some things improve with age like crystallized knowledge and, hopefully, wisdom. Life satisfaction and well-being also rise above age 60 on average.

The incidence of many diseases, called **age-related diseases**, also rises exponentially with age (as we will discuss in Chapter 8). Major age-related diseases include type-2 diabetes, heart failure, Alzheimer's disease, osteoarthritis, and most cancers. The incidence of many of these diseases rises with age with a similar slope of 6%–8% per year.

Another universal feature of aging is that the *variation between individuals increases with age* in most physiological functions. The young are typically similarly healthy, whereas the old can be healthy or sick to a wide range of degrees. The health of 20-year olds is like a mass-produced poster, whereas 80-year-olds are each an individually crafted work of art.

One way to quantify this variability is the **frailty index** (Mitnitski 2002). The frailty index is simple to define – the fraction of deficits a person has out of a list of deficits, ranging from back pain and hearing loss to diabetes and cancer. The frailty index can range between zero – no deficits, and one – all deficits on the list.

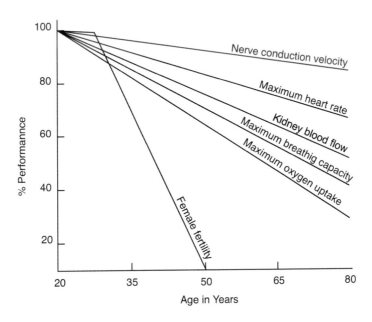

FIGURE 6.6   Most physiological functions decline with age. Adapted from Ginneken (2017).

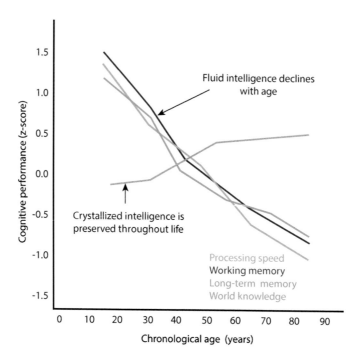

FIGURE 6.7 Most cognitive functions decline with age. Adapted from Zamroziewicz and Barbey (2018).

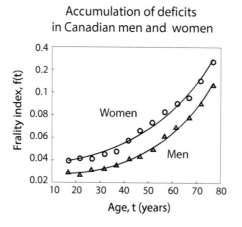

FIGURE 6.8 Frailty index – the fraction of health deficits an individual has from a list of deficits – accelerates with age. Adapted from Mitnitski et al. (2002).

The frailty index can help us understand in quantitative terms how health declines. The average frailty index increases in an accelerating way with age (Figure 6.8). The distribution of frailty becomes wider and skewed to high values with age (Figure 6.9A).

The standard deviation of frailty also grows with age. However, it grows more slowly than the mean. Therefore, the relative heterogeneity of frailty, the **coefficient of variation**

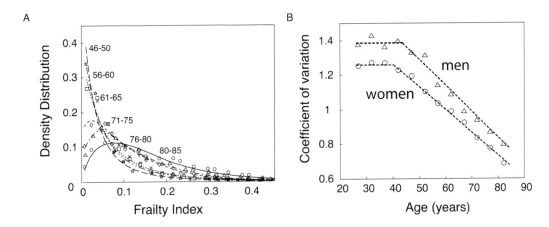

FIGURE 6.9 Distribution of frailty index is skewed to the right and shows decreasing coefficient of variation with age. Each curve is for a different age group. (A-B) described in text. Adapted from Rockwood, Mogilner, and Mitnitski (2004).

defined as the standard deviation/mean, goes down with age (Figure 6.9B). We will return to this point in the next chapter – the variation between individuals in frailty rises in absolute terms but drops in relative terms. There are differences in which deficits each person has, but in relative terms frailty becomes more similar between individuals with age.

This begins our survey of quantitative patterns of aging and aging-related decline – to set the stage for the next chapter that will explore organizing principles to explain these patterns. This may be a good moment for a nice deep sigh of relief.

GENETICALLY IDENTICAL ORGANISMS DIE AT DIFFERENT TIMES

Is the rate of decline due to the environment or genes? It turns out that the main effect is due to neither. Genetically identical organisms grown in the same conditions, such as identical twin lab mice, die at different times despite having the same genes and environment. Their relative variation in lifespan is about 30%, which is similar to the variation between unrelated mice. Such variation between genetically identical individuals is found in every organism studied, including flies and worms (Finch et al. 2000).

In humans as well, which are of course not genetically identical except in the case of identical twins, the heritable component of the variation in lifespan is small; more than 80% of the variation in lifespan is non-heritable. What is heritable is what people die of, as in genetic risks for cancer or diabetes.

The environment affects human mortality, of course. One important factor is low socioeconomic status that goes with higher risk of disease and death. A decade of lifespan separates the lowest and highest income deciles in many countries (Winkleby, Cubbin, and Ahn 2006). This disparity is found even when correcting for access to healthcare. It may in part be due to chronic stress accompanying low socioeconomic status.

Beyond these genetic and environmental factors, the evidence suggests that the risk of death in all organisms is dominated by *a large stochastic (random) component*.

LIFESPAN CAN BE EXTENDED IN MODEL ORGANISMS

At this point, it is important to say that the goal of most (credible) researchers studying aging is not to unlock the secrets of immortality, or even greatly extend human lifespan, but instead to understand the biological process of aging in order to extend the health span and reduce the burden of age-related disease. Lifespan data is, however, informative and exciting, and can help us to understand the fundamental drivers of aging.

Research on model organisms shows that lifespan can be extended. Certain mutations and interventions extend lifespan in roundworms up to three-fold, and in mice by up to 50%. A common factor for many such "longevity mutations" across different organisms is that they lie in pathways which control the tradeoff between growth and maintenance.

One such pathway is the IGF1 pathway. Mutants that inhibit this pathway turn on a starvation program that increases repair processes at the expense of growth. The mutant organisms thus grow more slowly and live longer. In humans, a mutation that disrupts the same pathway causes Laron dwarfism, which is associated with increased lifespan and decreased risk of cancer and type-2 diabetes.

Nutrition can also affect longevity, in part by acting through the same IGF1 pathway: continuous caloric restriction that reduces 30%–40% of normal calorie intake can extend lifespan in organisms ranging from yeast to monkeys. Variations on this theme also extend lifespan, such as restricting the time for feeding and restricting certain components of diet. In animals like flies and roundworms, lower temperature also increases lifespan.

The survival curves with these lifespan-changing perturbations show an extended mean lifetime, as seen by their shifted halfway point (Figure 6.10). But when time is rescaled by the average lifespan, the survival curves for most (but not all) perturbations line up with each other, showing that they have the same shape (Figure 6.11). This **scaling** property, discovered in *C. elegans* by Stroustrup et al. (2016) (see also Liu and Acar (2018)), suggests that the stochastic processes of aging may have a single dominant timescale that determines longevity.

What if the intervention for lifespan extension begins in mid-life? Interestingly, flies shifted from a normal diet to a lifespan-extending diet show rapid shifts to the better Gompertz curve within days (Mair et al. 2003). This suggests that there is a second, more rapid timescale to the stochastic process of aging (Figure 6.12). The same rapid shift also occurs the other way, when flies are shifted from life-span extending diet to normal diet.

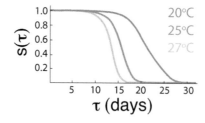

FIGURE 6.10 Survival curves for *C. elegans* show that median lifespan drops with temperature of growth. Adapted from Stroustrup et al. (2016).

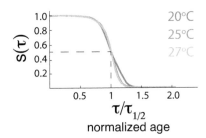

FIGURE 6.11 Survival curves for *C. elegans* fall on the same curve when age is normalized by the median lifespan. This property is known as survival curve scaling. Adapted from Stroustrup et al. (2016).

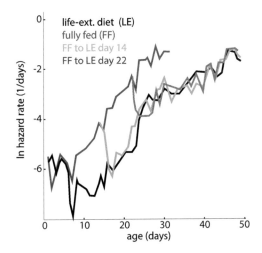

FIGURE 6.12 Fruit flies show rapid shifts to the better hazard curve upon a shift from normal to life-span extending diet. Adapted from Mair et al. (2003).

Other perturbations in flies, such as a temperature shift, show a change in Gompertz slope (Figure 6.13), but not a complete shift to a new curve altogether. In the next chapter, we will explain such dynamics.

LIFESPAN IS TUNED IN EVOLUTION ACCORDING TO DIFFERENT LIFE STRATEGIES

In contrast to the modest extension of lifespan in laboratory experiments, natural selection can tune lifespan by a factor of 100 between mammals, ranging from 2 years for shrews to 200 years for whales.

Aging rates thus evolve. Why does aging evolve? Early ideas were that aging is programmed because death offers a selective advantage at the population level. Get rid of old professors to allow space for new faculty. However, these theories don't generally seem to hold up in simulations.

Evolutionary theories of aging have converged on an idea called the **disposable soma theory** (Kirkwood 1977). This today dominates evolutionary thinking on aging. The theory notes that organisms wield a finite level of biological resources. They face a tradeoff

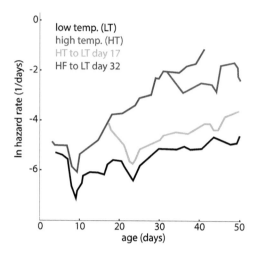

FIGURE 6.13 Fruit fly hazard curves change slope upon a shift in temperature. Adapted from Mair et al. (2003).

between repairing their bodies (soma) and reproducing. When they are subject to high predation, it's better for them to invest those resources in rapid growth and reproduction.

Thus, if an animal has high extrinsic mortality, like a mouse that is killed by predators within 1 year on average, it does not make sense to invest in repair processes that ensure a lifespan of 10 years. Instead, the mouse invests in growth and reproduction, making a lot of babies before extrinsic mortality finishes it off. In contrast, low extrinsic mortality as in elephants and whales selects for investment in repair, allowing a longer lifespan.

Indeed, large animals generally face less predation than small animals and live longer. A well-known relation connects mass to longevity: on average, longevity follows the fourth root of mass, $L \sim M^{\frac{1}{4}}$. A 100-ton whale is 10^8 heavier than a 1 g shrew, and thus should live 100 times longer, matching their 200 year versus 2-year lifespans.

However, there are exceptions. Bats weigh a few grams, like mice, but live for 40 years, which is 20 times longer than mice; similarly, naked mole rats weigh 10 g and live for decades. Pablo Szekely, in his PhD with me, plotted longevity versus mass for all mammals and birds for which data was available (Szekely et al. 2015). Instead of a line, the data falls inside a wedge-shaped distribution that we called the mass-longevity triangle (Figure 6.14).

At the vertices of the triangle are shrews, whales, and bats. These three vertices represent three archetypal life strategies. Shrews and mice have a **live fast die young** strategy, as described above. Whales and elephants, in contrast, have very low predation due to their enormous size. They have a **slow life strategy** of producing a single offspring at a time and caring for it for a long time. Bats have a protected niche (flying) and thus, despite their small size, they face low predation. The **protected niche strategy** entails the longest childhood training relative to lifespan and the largest brain relative to body mass. Bats carry a baby on their back to teach it, for example, where specific fruit trees are found.

In the triangle near the bats are other animals with protected niches, such as tree-living squirrels, the naked mole rat that lives underground, primates with their cognitive niche, and flying (as opposed to flightless) birds. Flightless birds have shorter lifespan than flying birds of the same mass and lie closer to the bottom edge of the triangle.

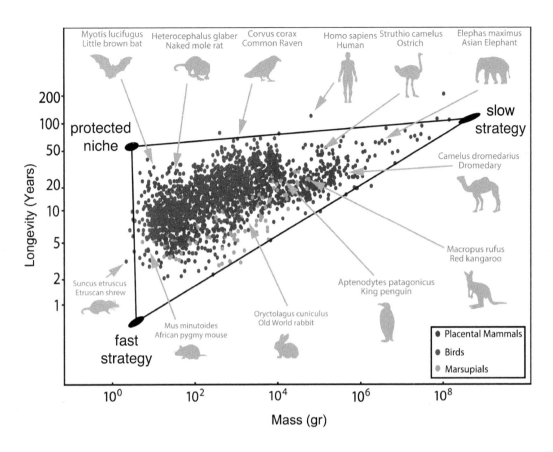

FIGURE 6.14 The mass-longevity triangle. Each dot is a mammalian or bird species. The vertices correspond to archetypal life-strategies denoted shrew (S), whale (W), and bat (B). Adapted from Szekely et al. (2015).

Why the triangular shape? Why are there no mammals below the triangle, namely large animals with short lives? It takes time to build a large mass, and thus such animals may be unfeasible. An additional answer is provided by the theory of **multi-objective optimality** in evolution. Tradeoff between three strategies, according to this theory, results in a triangle shape in trait space (Shoval et al. 2012). The triangle is the set of all points that are closest to the three vertices, which represent archetypal strategies. The closer a point is to a vertex, the better it performs the vertex strategy. For any point outside the triangle, there is a point inside that is closer to all three vertices and is thus more optimal (Shoval et al. 2012; Szekely et al. 2015). Phylogenetic relatedness on its own does not explain this triangle shape, because species from very different families often lie close to each other on the triangle (Adler et al. 2022).

All in all, bigger species tend to live longer. But above we mentioned that within a species, there is an opposite trend – bigger individuals are shorter lived than smaller ones, such as the IGF1 mutants described above. Longevity and mass *within* a species often go against the trend seen *between* species. In dogs, for example, tiny Chihuahuas live 15–20 years whereas Great Danes live for 4–6 years. Some of the mutations that occurred during the breeding of these dogs are in the IGF1 pathway. Evidently, natural selection tunes longevity in different species by other means than adjusting their IGF1 pathway. Current evidence points instead to increased repair capacity in long-lived species.

So far, we discussed the **population statistics** of aging. Such work requires counting deaths. What about the molecular mechanisms of aging? Molecular causes of aging are intensely studied. However, the molecular study of aging and the population study of aging are two disciplines that are rarely connected. Our goal, in the next chapter, *will be to bridge the molecular level and the population level laws of aging.* To do so, we need to first discuss the molecular causes of aging.

MOLECULAR THEORIES OF AGING FOCUS ON CELLULAR DAMAGE

There are several molecular theories of aging, each focusing on a particular kind of damage to the cell and its components. The main types of cellular damage include DNA damage, protein damage, and damage to the cells' membranes or their energy factories called mitochondria. An important cause of such damage is reactive oxygen species (ROS), leading to the ROS theory of aging. Each molecular theory arose because disrupting a repair mechanism that fixes a specific kind of damage causes accelerated aging. For example, disrupting certain types of DNA repair enzymes, such as certain helicases, causes accelerated aging both in model organisms and in humans in rare genetic diseases that cause premature aging, such as Werner's syndrome. Likewise, disrupting repair processes that dispose of unfolded proteins or damaged mitochondria, called autophagy and mitophagy, can cause premature aging.

Another theory of aging is based on the fact that with each cell division, the ends of the DNA chromosomes called *telomeres* become shorter. When telomeres become too short, the cell can no longer divide. Thus, telomeres limit the number of cell divisions. Indeed, average telomere length drops with chronological age and drops faster in some conditions of accelerated aging.

These theories have been collected into hallmarks of aging (Figure 6.15; López-Otín et al. 2023). Each hallmark ideally satisfies three criteria: it rises with age, enhancing it speeds aging, and attenuating it slows aging in model organisms.

In the past, researchers have sought for a single factor that drives aging. Currently, there is much confusion about whether such a single factor exists; the field generally has come to view aging as an emergent property of many interacting factors. There is a need for simplifying concepts (Cohen et al. 2022).

None of the aging hallmarks have been connected to the Gompertz law or the other quantitative patterns of aging discussed above. Making this connection is the goal of the next chapter. To prepare, we first need to explore what kind of damage can accumulate over decades.

DNA ALTERATIONS IN STEM CELLS CAN ACCUMULATE FOR DECADES

To make progress, we now enter the frontier of research. We saw that aging means that something in our body changes over decades leading to dysfunction. Let's ask what fundamental aspects are required for damage to accumulate with age. As we discussed in

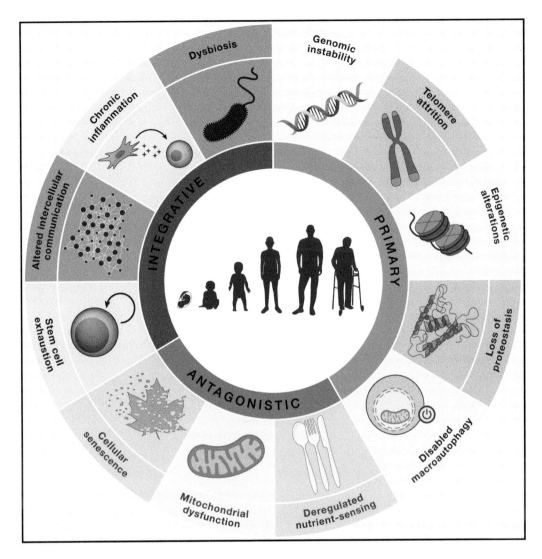

FIGURE 6.15 Hallmarks of aging are factors that rise with age, whose enhancement speeds aging and whose attenuation slows it down. Adapted from López-Otín et al. (2023).

Chapters 1–3, many tissues have cells that turn over within weeks to months. If one of these cells becomes damaged, it will be removed within months. That kind of damage doesn't accumulate over decades.

In order to accumulate over decades, the source of damage must remain in the body permanently. Therefore, the source of damage that we care about should be transmitted in cells *that are not removed*. Since all organs age, these cells should be found throughout the body.

A good candidate for such ubiquitous and persistent cells is **stem cells.** To understand stem cells, let's consider the skin as an example. The top layer of the skin is made of dead cells that are removed within weeks. To make new skin cells, a deep skin layer called the basal layer of the epidermis houses skin stem cells, *S* (Figure 6.16).

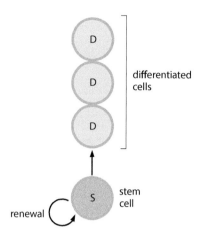

FIGURE 6.16 Stem cells S divide and differentiate into differentiated cells D. Stem cells form a lineage that often remain in the body for a lifetime, whereas differentiated cells in most tissues are removed in days to years.

These stem cells divide to make new stem cells, in a process called **stem cell renewal**. They also **differentiate** into skin cells, D. These differentiated skin cells divide only a few times, a process called transit amplification. Each stem cell division gives rise to many differentiated cells due to transit amplification. The cells rise in a column above the stem cells, until they reach the top layer of the skin, and are shed off. The stem cells continuously and slowly divide to replace the lost skin cells. They have enzymes that replenish their telomeres so they have virtually unlimited division potential.

Many tissues have their own dedicated stem cells. Stem cells are found for example in the epithelial lining of the intestine, lungs and skin. Stem cells in the bone marrow differentiate about once per month to produce the red and white blood cells.

Since stem cells stay in the body throughout life, and all cells mutate, they run the risk of gaining mutations and other changes in their DNA. Stem cells gain on the order of 50 mutations per year in humans due to passive chemical damage to their DNA. They gain a further few mutations with each division. Most of these mutations do nothing. A few are harmful to the stem cell, making it die or grow slower than its neighbor stem cells, and therefore such mutant stem cells are lost.

But some mutations lead to changes in genes that don't harm the stem cells but affect proteins expressed in its progeny, the differentiated cells, D. These mutations encode for production of malfunctioning proteins that cause cellular damage in the differentiated cells. For example, the malfunctioning protein might mis-fold and gum up the differentiated cell, or produce ROS, which damages the DNA and proteins of the differentiated cell.

Thus, with age, there will be more and more mutant stem cells, denoted S', that produce damaged differentiated cells, D' (Figure 6.17). Since the mutations don't affect these stem cells, they are "invisible" and the immune system cannot remove them. Above each such mutant, stem cell will be a column of damaged cells. The number of these "damaged-cell factories" increases with age.

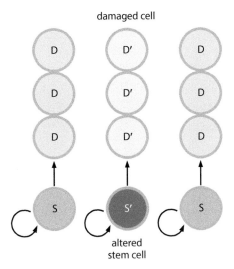

FIGURE 6.17 Stem cells with certain alterations can generate damaged differentiated cells, with no harm to the altered stem cell itself. Such altered stem cells remain indefinitely in the body and act as damage producing units.

Indeed, measurements by Stratton and colleagues (Figure 6.18) found that the number of mutations on average in each human stem cell rises linearly with age, reaching about 3000 point mutations by age 60. Therefore, the number of mutant stem cells S' that happen to have a dangerous mutation for the differentiated cells should also rise linearly with age, $S' \sim \tau$. In other words, the number of mutant cells tracks chronological time.

Interestingly, non-dividing neurons have a similar number of mutations and may represent another repository of damage that stays in the body for a lifetime.

Strikingly, mice reach a similar number of mutations by the age of 2 years. Dogs reach this by about 12 years. Different mammalian species have an effective rate of mutation accumulation that scales as 1/lifespan (Cagan et al. 2022) – the shorter the lifespan the faster mutations accumulate.

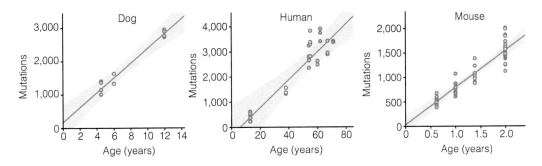

FIGURE 6.18 The number of mutations per crypt stem cell rises linearly with age in different species and reaches several thousand at old age. Adapted from Cagan et al. (2022).

Mutations alone, however, do not seem to account for aging. Humans and mice with enlarged mutation rates, such as those with defects in DNA polymerase or mismatch repair, do not generally show premature aging (Robinson et al. 2021).

A more relevant DNA alteration that rises with chronological age is **epigenetic changes** to the DNA. For example, some sites in the DNA become methylated or unmethylated, in a process that is stochastic and has a low rate. These methylations can alter gene expression. Other epigenetic changes include histone acetylation. They open up normally silenced regions such as regions near telomeres and cause aberrant transcription and even DNA damage due to structures in which RNA binds DNA improperly called *R*-loops. Many such epigenetic changes rise linearly with age during adulthood, giving rise to "aging clocks" (Xia et al. 2021; Horvath and Raj 2018). Finally, the human genome contains numerous virus genomes called retrotransposons which can jump into new genomic regions and disrupt genes. The number of such jumps also rises with age (Andrenacci, Cavaliere, and Lattanzi 2020), and is another source of mutated stem cells.

There are other ways that stem cells can accumulate damage that they pass on to differentiated cells. For example, protein translation sometimes makes errors, and the misfolded proteins can have a dangerous conformation that causes them to form large toxic aggregates such as polymers or gels. This is a contagious sort of conformation, called a prion state - once it occurs, it is hard to get rid of because it causes properly folded proteins to assume the same prion state and aggregate too. Prion states are thus a type of memory that can last for decades. Again, since translation errors are rare, the probability of having a prion state in a given stem cell is also expected to rise slowly and proportionally to age.

We can treat these different types of stem cell alterations similarly for our purposes. We will call cells that produce damaged progeny **altered stem cells**. The main point is that the number of such altered stem cells rises approximately linearly with age throughout adulthood.

DAMAGED AND SENESCENT CELLS BRIDGE BETWEEN MOLECULAR DAMAGE AND TISSUE-LEVEL DAMAGE

What happens to the damaged cells, D'? As we saw in Chapter 5 on wound healing, damaged cells send signals that call in the immune system, generating inflammation. Sometimes, damaged cells commit programmed cell death, apoptosis, a process in which cells quickly and cleanly remove themselves. Damaged cells, however, often take another route: they become zombie-like **senescent cells (SnC)** (Figure 6.18).

Senescent cells serve an essential purpose in young organisms: they guide the healing of injury. When organisms are injured, cells sense that they have been damaged. If they keep dividing, they run the risk of becoming cancer cells. However, if all injured cells kill themselves, the tissue will have a hole. Therefore, turning into a senescent cell maintains tissue integrity without the risk of cancer.

Here we focus on senescent cells as a plausible driver of aging.

Senescent cells are large and metabolically active cells which secrete signal molecules, collectively called the **senescence-associated secretory phenotype**, or **SASP**

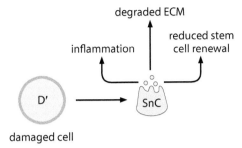

FIGURE 6.19   Damaged cells that are unable to repair themselves become senescent cells. Senescent cells secrete proteins called SASP that cause inflammation, degrade the extracellular matrix, and reduce stem cell renewal.

(Figure 6.19). The SASP includes signaling molecules that recruit the immune system to clear the senescent cells in an organized fashion. In other words, these signals cause inflammation. Certain cells of the immune system are tasked with detecting and removing senescent cells, such as macrophages and NK cells. The NK cells and macrophages also have other important jobs such as removing virus-infected cells and cancer cells.

SASP also slows down the rate of stem-cell renewal around the senescent cells, to wait for the orderly clearance of the senescent cells by the immune system. Finally, SASP contains "molecular scissors" that alter the extracellular matrix (ECM) around the cells, to allow the immune system to enter.

Thus, after an injury, senescent cells arise. They cause inflammation to call in the immune cells that remove them in an orderly process over several days to allow healing.

However, senescent cells have a dark side. This dark side arises because we are not designed to be old. As we age, alterations accumulate in stem cells. The altered stem cells S' produce damaged cells D', which "think" that there is an injury. Some of these damaged cells turn into senescent cells. The number of such altered stem cells, or "damage producing units," rises linearly with age as we saw. As a result, the production rate of senescent cells rises with age, production $\sim \eta\tau$ throughout the body. Eventually, according to our law 2, their removal processes saturate. When production exceeds removal, all bets are off and senescent cell abundance skyrockets. In the next chapter, we will understand the ramifications of saturation of removal capacity.

Because the aging body becomes loaded with damaged and senescent cells, it is permeated with SASP. Even if, say, only 0.1% of the cells are damaged, their secreted inflammatory signals can affect the entire body, causing chronic inflammation. This is a hallmark of aging, sometimes called **inflammaging**. Inflammaging drives nearly all age-related diseases including osteoarthritis, diabetes, Alzheimer's disease and heart disease. The SASP also slows stem cell renewal all over the body and alters the extracellular matrix. These effects increasingly lead to reduction in organ function.

Thus, senescent cells sit at an interesting junction between the level of damage to cell components and the level of damage to organ systems (Figure 6.20). They unite the different

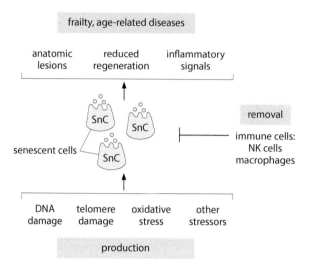

FIGURE 6.20 Almost all types of cellular damage can lead to senescent cells, which produce SASP proteins that lead to systemic effects that cause age-related diseases. Senescent cells are removed by immune cells including NK cells and macrophages.

molecular theories of aging, because virtually any form of cellular damage can cause a cell to become senescent, including ROS, DNA damage, shortened telomeres, epigenetic damage, and so on. And senescent cells in turn can produce systemic effects that cause disease and physiological decline.

REMOVING SENESCENT CELLS IN MICE SLOWS AGE-RELATED DISEASES AND INCREASES AVERAGE LIFESPAN

In 2016, an experiment by van Duersen et al. (Baker et al. 2016) galvanized the aging field. It showed that accumulation of senescent cells is causal for aging in mice: continuous targeted elimination of senescent cells increased mean lifespan by 25%. Such removal also attenuated the age-related deterioration of heart, kidney, and fat. The original 2016 results have been reproduced by other research groups using different methods to remove senescent cells. These methods include drugs called **senolytics** that selectively kill senescent cells in mice. There are several classes of senolytic drugs. Senolytics delay cancer development and ameliorate age-related diseases including diabetes, osteoarthritis, Alzheimer's, and heart disease in mice models.

For a sense of the effects of senescent cell removal, see the picture of twin mice at age 2 years (Figure 6.20), roughly equivalent to age 70 in humans. One sibling had senescent cells removed continually starting at the age of 1 year. It ran on the wheel and had shiny fur and overall better health. Its sibling, treated with mock injections, barely ran on the wheel and looked like a typical aged mouse with a hunched back, cataract, and fur loss (Figure 6.21).

FIGURE 6.21 Rejuvenation of an old mouse genetically engineered to remove its senescent cells (p16 expressing cells) compared to its littermate in which senescent cells were not removed. Adapted from Baker et al. (2016).

Accumulation of senescent cells is not the only cause of aging, as evidenced by the fact that these mice still age, get sick, and die. But in the next chapter, we will assume that they are a dominant cause. We will also make the simplifying assumption that senescent cells are a single entity, even though they are heterogeneous and tissue-specific. These simplifying assumptions will help us write a stochastic process that can explain many of the empirical aging patterns that we have seen.

EXERCISES

Solved Exercise 6.1: Survival and hazard
Survival $S(\tau)$ is the probability of dying after age tau. Hazard $h(\tau)$ is the probability per unit time to die.

a. Show that $h(\tau) = -\dfrac{1}{S(\tau)}\dfrac{dS}{d\tau}$.

Solution: consider a cohort of N_o individual born at the same time. The number that survive until at least age τ is $N(\tau) = N_o\, S(\tau)$. The number $D(\tau)$ that die in a small time interval δt around age τ is the number that survived till τ but not till $\tau + \delta t$: $D(\tau) = N_o S(\tau) - N_o S(\tau + \delta t)$. When δt is small, this equals $D(\tau) = -N_o \dfrac{dS}{d\tau}\delta t$. The minus sign is due to the fact that survival S decreases, and thus has a negative slope. The hazard $h(\tau)$ is the probability per unit time to die for organisms at age τ, and thus $h(\tau)\delta t = D(\tau)/N(\tau)$, so that $h(\tau) = -\dfrac{1}{S(\tau)}\dfrac{dS}{d\tau}$.

b. Show that this means that

(P6.1)
$$S(\tau)=e^{-\int_0^\tau h(t)dt}$$

Exercise 6.2:
Use Equation P6.1 to solve and plot the survival curve in the following cases:

a. Constant hazard $h = h_0$

b. Linearly rising hazard: $h(\tau) = \alpha\tau$

c. Gompertz law: $h(\tau) = b\,e^{\alpha\tau}$. In humans, α is about 0.085 $year^{-1}$, implying a doubling of mortality every $log(2)/\alpha = 0.69/0.085 = 8$ years.

d. Trees with hazard that drops with age as $h = a/(1+b\tau)$.

e. Gompertz–Makeham law, in which age-independent extrinsic mortality is added: $h(\tau) = be^{\alpha\tau} + h_0$. Estimate the parameters for human data based on Figures 6.4 and 6.5.

Exercise 6.3:
What is the median half-life in each of the cases of Solved Exercise 6.2, defined as that age $\tau_{1/2}$ at which $S(\tau_{1/2}) = 0.5$?

Exercise 6.4: Gompertz law with slowdown: one relation that models the slowdown in hazard at old ages is called the Gamma–Gompertz law: $h(\tau) = a\dfrac{exp(\alpha\tau)}{1+b\,exp(\alpha\tau)}$.

a. What is the survival curve $S(\tau)$?

b. What is the median lifespan?

c. Estimate (roughly) the parameters a, b, and α that describe intrinsic mortality of human data in Figures 6.4 and 6.5. What is the estimated human median lifespan?

Exercise 6.5: Lifespan distribution
What is the distribution of lifespans given a hazard curve $h(\tau)$?

Exercise 6.6: Maximal lifespan
Consider a population of N individuals with a survival curve $S(\tau)$.

a. Why can the maximal lifespan be roughly estimated as the age τ when $S(\tau) = 1/N$?

b. What is the estimated maximal lifespan for the case of the Gompertz law? How does it depend on population size?

c. The world's population is about $N \sim 10^{10}$ people. Use the estimate of hazard from Exercise 6.4 to predict the maximal lifespan if there were 10^9 people, or 10^{11}, if all parameters remain the same. Note for reference that the longest human lifespan is currently thought to be of a woman who died at 122.

Exercise 6.7: Disposable soma theory

a. Use evolutionary thinking to explain the phenomenon of menopause, which happens in a few species including humans and elephants.

b. A gene has "antagonistic pleiotropy," meaning that it provides reproductive advantage to a young reproductive organism but reduces survival at old age. How would natural selection affect the gene's frequency in the population?

c. Consider the case of senescent cells. What type of biological mechanism such as production or removal of senescent cells can serve as a possible place to look for antagonistic pleiotropy?

Exercise 6.8: Mass-longevity triangle
Look up Figure 3 in the paper on the mass-longevity triangle (Szekely et al 2016). This figure shows various life-history features of animals, relative to the mean, as a function of their distance on the triangle from the three vertices. For example, panel a shows litter size (number of babies per birth), with the animals closest to the shrew (S), bat (B), or whale (W) vertex at $x=0$. Stars indicate statistically significant increase or decrease in the animals closest to the vertex. Choose four features and provide a brief explanation of these trends in terms of life strategies.

Exercise 6.9: Estimated longevity: In the mass-longevity triangle, longevity is the maximal lifespan observed for each species, based on the Anage database (Tacutu et al. 2018). Discuss possible sources of error in this estimated maximal lifespan. How would such errors affect the shape of the distribution of longevity versus mass?

Exercise 6.10: Strehler–Mildvan correlation as an artifact
Read the following two papers and discuss a correlation in Gompertz law parameters that may result from a fitting artifact (Finkelstein 2012; Tarkhov, Menshikov, and Fedichev 2017).

Exercise 6.11: Explain the difference in the impact of mutations in stem cells and in germ cells.
Germ cells accumulate ~100 mutations between generations but undergo strong quality control and negative selection that removes mutants of strong effect. Germs cells have most epigenetic marks removed, unlike somatic cells. What is the impact of these features on aging?

REFERENCES

Adler, Miri, Avichai Tendler, Jean Hausser, Yael Korem, Pablo Szekely, Noa Bossel, Yuval Hart, et al. 2022. "Controls for Phylogeny and Robust Analysis in Pareto Task Inference." *Molecular Biology and Evolution* 39 (1): msab297. https://doi.org/10.1093/molbev/msab297.

Andrenacci, Davide, Valeria Cavaliere, and Giovanna Lattanzi. 2020. "The Role of Transposable Elements Activity in Aging and Their Possible Involvement in Laminopathic Diseases." *Ageing Research Reviews* 57:100995. https://doi.org/10.1016/j.arr.2019.100995.

Baker, Darren J., Bennett G. Childs, Matej Durik, Melinde E. Wijers, Cynthia J. Sieben, Jian Zhong, Rachel A. Saltness, et al. 2016. "Naturally Occurring p16^{Ink4a}-Positive Cells Shorten Healthy Lifespan." *Nature* 530: 184–9. https://doi.org/10.1038/nature16932.

Barbieri, Magali, John R. Wilmoth, Vladimir M. Shkolnikov, Dana Glei, Domantas Jasilionis, Dmitri Jdanov, Carl Boe, et al. 2015. "Data Resource Profile: The Human Mortality Database (HMD)." *International Journal of Epidemiology* 44 (5): 1549–56. https://doi.org/10.1093/ije/dyv105.

Cagan, Alex, Adrian Baez-Ortega, Natalia Brzozowska, Federico Abascal, Tim H. H. Coorens, Mathijs A. Sanders, Andrew R. J. Lawson, et al. 2022. "Somatic Mutation Rates Scale with Lifespan across Mammals." *Nature* 604 (7906): 517–24. https://doi.org/10.1038/s41586-022-04618-z.

Carnes, Bruce A., Larry R. Holden, S. Jay Olshansky, M. Tarynn Witten, and Jacob S. Siegel. 2006. "Mortality Partitions and Their Relevance to Research on Senescence." *Biogerontology* 7 (4): 183–98. https://doi.org/10.1007/s10522-006-9020-3.

Cohen AA, Ferrucci L, Fülöp T, Gravel D, Hao N, Kriete A, Levine ME, Lipsitz LA, Olde Rikkert MG, Rutenberg A, Stroustrup N, Varadhan R. 2022. "A complex systems approach to aging biology." *Nature Aging* 2(7): 1-12. DOI: 10.1038/s43587-022-00252-6

Finch, Caleb Ellicott, Arco and William Kieschnick Professor of Gerontology and Director of the Alzheimer Research Center Caleb E. Finch, T. B. L. Kirkwood, Tom Kirkwood, Professor and Director of the Joint Center on Aging Tom Kirkwood, and Professor of Medicine and Head of the Department of Gerontology Tom Kirkwood. 2000. *Chance, Development, and Aging.* Oxford University Press.

Finkelstein, Maxim. 2012. "Discussing the Strehler-Mildvan Model of Mortality." *Demographic Research* 26: 191–206. https://doi.org/10.4054/DemRes.2012.26.9.

van Ginneken, Vincent. 2017. "Are There Any Biomarkers of Aging? Biomarkers of the Brain." *Biomedical Journal of Scientific & Technical Research* 1 (1): 2017.

Horvath, Steve, and Kenneth Raj. 2018. "DNA Methylation-Based Biomarkers and the Epigenetic Clock Theory of Ageing." *Nature Reviews Genetics* 19 (6): 371–84. https://doi.org/10.1038/s41576-018-0004-3.

Kirkwood, T. 1977. "Evolution of ageing". *Nature.* **270** (5635): 301–304.

Liu, Ping, and Murat Acar. 2018. "The Generational Scalability of Single-Cell Replicative Aging." *Science Advances* 4 (1): eaao4666. https://doi.org/10.1126/sciadv.aao4666.

López-Otín, Carlos, Maria A. Blasco, Linda Partridge, Manuel Serrano, and Guido Kroemer. 2023. "Hallmarks of Aging: An Expanding Universe." *Cell.* 186(2):243-278. doi: 10.1016/j.cell.2022.11.001. PMID: 36599349.

Mair, William, Patrick Goymer, Scott D. Pletcher, and Linda Partridge. 2003. "Demography of Dietary Restriction and Death in Drosophila." *Science* 301 (5640): 1731–33. https://doi.org/10.1126/science.1086016.

Mitnitski, Arnold B., Alexander J. Mogilner, Chris MacKnight, and Kenneth Rockwood. 2002. "The Accumulation of Deficits with Age and Possible Invariants of Aging." *TheScientificWorldJournal* 2:573409. https://doi.org/10.1100/tsw.2002.861.

Miwa, Satomi, Sonu Kashyap, Eduardo Chini, and Thomas von Zglinicki. n.d. "Mitochondrial Dysfunction in Cell Senescence and Aging." *The Journal of Clinical Investigation* 132 (13): e158447. https://doi.org/10.1172/JCI158447.

Robinson, Philip S., Tim H. H. Coorens, Claire Palles, Emily Mitchell, Federico Abascal, Sigurgeir Olafsson, Bernard C. H. Lee, et al. 2021. "Increased Somatic Mutation Burdens in Normal Human Cells Due to Defective DNA Polymerases." *Nature Genetics* 53 (10): 1434–42. https://doi.org/10.1038/s41588-021-00930-y.

Rockwood, K., A. Mogilner, and A. Mitnitski. 2004. "Changes with Age in the Distribution of a Frailty Index." *Mechanisms of Ageing and Development* 125 (7): 517–519.

Shoval, O., H. Sheftel, G. Shinar, Y. Hart, O. Ramote, A. Mayo, E. Dekel, et al. 2012. "Evolutionary Trade-Offs, Pareto Optimality, and the Geometry of Phenotype Space." *Science* 336 (6085): 1157–60. https://doi.org/10.1126/science.1217405.

Stroustrup, Nicholas, Winston E. Anthony, Zachary M. Nash, Vivek Gowda, Adam Gomez, Isaac F. López-Moyado, Javier Apfeld, et al. 2016. "The Temporal Scaling of Caenorhabditis Elegans Ageing." *Nature* 530:103–7. https://doi.org/10.1038/nature16550.

Szekely, P., Y. Korem, U. Moran, A. Mayo, and U. Alon. 2015. "The Mass-Longevity Triangle: Pareto Optimality and the Geometry of Life-History Trait Space." *PLoS Computational Biology* 11 (10): e1004524. https://doi.org/10.1371/journal.pcbi.1004524.

Tacutu, R., D. Thornton, E. Johnson, A. Budovsky, D. Barardo, T. Craig, ..., & J.P. De Magalhães. 2018. "Human Ageing Genomic Resources: New and Updated Databases." *Nucleic Acids Research* 46(D1): D1083–D1090.

Tarkhov, Andrei E., Leonid I. Menshikov, and Peter O. Fedichev. 2017. "Strehler-Mildvan Correlation Is a Degenerate Manifold of Gompertz Fit." *Journal of Theoretical Biology* 416:180–9. https://doi.org/10.1016/j.jtbi.2017.01.017.

Winkleby, Marilyn, Catherine Cubbin, and David Ahn. 2006. "Effect of Cross-Level Interaction between Individual and Neighborhood Socioeconomic Status on Adult Mortality Rates." *American Journal of Public Health* 96 (12): 2145–53. https://doi.org/10.2105/AJPH.2004.060970.

Xia, Xian, Yiyang Wang, Zhengqing Yu, Jiawei Chen, and Jing-Dong J. Han. 2021. "Assessing the Rate of Aging to Monitor Aging Itself." *Ageing Research Reviews* 69:101350. https://doi.org/10.1016/j.arr.2021.101350.

Yang, Yifan, Ana L. Santos, Luping Xu, Chantal Lotton, François Taddei, and Ariel B. Lindner. 2019. "Temporal Scaling of Aging as an Adaptive Strategy of Escherichia Coli." *Science Advances* 5 (5): eaaw2069. https://doi.org/10.1126/sciadv.aaw2069.

Zamroziewicz, Marta K., and Aron K. Barbey. 2018. "Chapter 2 – The Mediterranean Diet and Healthy Brain Aging: Innovations from Nutritional Cognitive Neuroscience." In *Role of the Mediterranean Diet in the Brain and Neurodegenerative Diseases*, edited by Tahira Farooqui and Akhlaq A. Farooqui, 17–33. Academic Press. https://doi.org/10.1016/B978-0-12-811959-4.00002-X.

CHAPTER 7

Aging and Saturated Repair

W E ARE NOW READY TO BUILD A THEORY OF AGING BASED on the patterns and molecular mechanisms we have seen. Our payoff will be a first-principle explanation of why genetically identical organisms die at different times. We will also understand the origin of the Gompertz law and the dynamics of anti-aging interventions.

A THEORY FOR AGING BASED ON SATURATING REMOVAL OF DAMAGE

To link the dynamics of aging to molecular and cellular mechanisms, we present a theory based on our three laws. We will then test it in multiple ways. The theory is agnostic as to the form of damage that is causal for aging, and we will apply it to humans and to other organisms.

To get intuition about the model, let's begin with a story about trucks. A young organism is like a small village where each house produces garbage. The village has 100 garbage trucks, more than enough to clear the garbage. Every year a new house is built and with time the village grows. It eventually becomes a big city that produces a lot of garbage every day. However, this village was not designed to be so large, so there are still only 100 trucks. But now the trucks are overloaded, and garbage piles up in the streets. Any extra garbage stays around for a long time until the overwhelmed trucks get to it. Eventually, when garbage is produced at a rate larger than the maximal capacity of the trucks, garbage piles up higher and higher. The situation becomes incompatible with life.

Let's interpret the story in terms of aging. The garbage is damage casual for aging, which we will call X. To be concrete, for mammals we assume that X is the total number of senescent cells in the body. These senescent cells secrete factors that cause chronic inflammation and reduced regeneration, leading to disease and decline. Later, when we discuss other organisms, X will represent other forms of damage.

The houses in the story are **damage producing units**, DPUs, that generate the damage X. They are produced at a constant rate and slowly accumulate in the body over decades. They make 70-year-olds different from 20-year-olds. Since DPUs are made at a constant rate but are irremovable, their number rises linearly with age. Therefore, the production rate of X rises linearly with age.

DOI: 10.1201/9781003356929-13

Our candidate for DPUs in mammals is altered stem cells. The alterations do not affect the stem cell but do affect the differentiated cells it produces. As we saw in Chapter 6, the number of altered stem cells rises linearly with age. They produce damaged differentiated cells that become senescent.

Houses, or DPUs, are not removed, but the damage that they create, X, is. It is removed by the trucks, which represent damage-removing processes. For senescent cells, the trucks are NK cells and macrophages. The NK cells detect senescent cells by means of special marker proteins that senescent cells display on their surface. The NK cells attach to the senescent cells and inject toxic proteins to kill them. Mice without functioning NK cells show accelerated aging and large amounts of senescent cells. Other immune cells, including macrophages, also play a role by swallowing up the remains of the killed cells.

But since all biological processes saturate (law 2), so does the removal capacity of senescent cells; the trucks become overwhelmed.

Thus, our model has two features: production of damage that rises linearly with age and saturating removal of damage (Figure 7.1).

This sets up an inevitable catastrophe – when production exceeds maximal removal capacity, the amount of damage X rises sharply.

To get a graphic sense of how damage X grows with age, we can use a rate plot. Removal rate saturates as a function of X, whereas production rate, represented by the colored horizontal lines, is low in young organisms and rises with age. The steady-state point is where the lines cross, as production equals removal. With age, the steady-state X accelerates to higher and higher levels (Figure 7.2) because of the saturating shape of the removal curve. When production rises above the removal curve, which occurs when age exceeds a critical age, all bets are off. The steady-state point shifts to infinity, and X grows indefinitely.

Indeed, with age, senescent cells pile up. They secrete factors that cause inflammation and reduce regeneration – their impact thus extends to the entire body. In order to have such systemic effects, the number of senescent cells does not have to be very large. For example, the entire body is affected by hormones like cortisol secreted by a 10g gland, which makes up less than 0.1% of the body's cells. Similarly, it may be enough to have only one out of a thousand cells become senescent for their secreted factors to affect physiology at large.

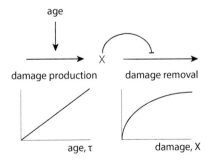

FIGURE 7.1   The saturating removal model for aging is based on damage that saturates its own removal and whose production rate rises linearly with age.

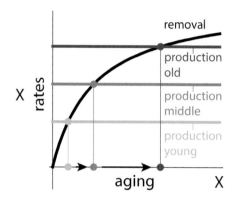

FIGURE 7.2   A rate plot shows that steady-state damage, the point at which production and removal curves intersect, accelerates with age.

The saturation of the immune cell "garbage trucks" contributes to decline in another important way: saturation hampers their other tasks, including fighting infection and cancer. Together, these systemic changes increase the risk of illness and organ dysfunction.

Another essential point in the theory is separation of timescales. Just as garbage is made and removed daily, much faster than the rate at which houses are built, so is damage X made and removed much faster than the slow accumulation of damage-producing units. Senescent-cell half-life is days to weeks, whereas their production rate, given by the number of altered stem cells, rises over decades.

Why doesn't the number of trucks rise with age? We can think about this from the point of view of natural selection, in terms of the disposable soma theory. Damage removal capacity, the number of trucks, is selected for the young. The amount of NK cells and macrophages is designed to allow young organisms to fight infection and recover from injury. We are not designed to be old – natural selection does not support a mechanism to sufficiently increase removal capacity at old age since most individuals in the wild never reach old age. Indeed, the number of NK cells does not increase strongly with age, they just seem more exhausted (Brauning et al. 2022).

More precisely, we can discount as our X any form of damage whose repair mechanism rises with age in a way sufficient to effectively remove X. Damage casual for aging in this theory needs to saturate its trucks when its levels rise sufficiently.

It is likely that each organism has several types of houses, generating several types of garbage, each with its own kind of truck. Perhaps altered stem cells and senescent cells are the first to accumulate; if they were to be eliminated, we might expose the next set of houses and garbage, perhaps damaged neurons, and so on.

In a nutshell, our theory for aging derives from the three laws:

- All cells come from cells – Stem cells produce differentiated cells.

- Cells mutate – Epigenetically altered stem cells increase linearly with age, and produce damaged and senescent cells.

- Biological processes saturate – the removal processes of these damaged and senescent cells eventually reach their maximal capacity and saturate. Damaged/senescent cell levels rise sharply leading to inflammaging and decline.

Natural selection is the driving force, according to disposable soma theory, because damage removal capacity was selected for the young, not the old, and hence does not increase with age. A useful physiological process in the young, production of senescent cells that cause inflammation as part of normal injury repair, is pushed beyond its design specifications to cause inflammaging.

SENESCENT-CELL DYNAMICS IN MICE CAN TEST THE MODEL

Our next step is to see what this model predicts for damage, X, which in mammals we assume is senescent cell accumulation with age. We would like to compare this to experimental data.

To get a feeling for the dynamics of senescent cells, let's consider an experiment by (Burd et al. 2013) who measured senescent-cell abundance in 33 mice every 8 weeks for 80 weeks. To measure whole-body senescent cell amounts, they used genetic engineering to produce mice that emit photons in proportion to the number of senescent cells (Figure 7.3). Photons are produced by a gene from fireflies called **luciferase** that produces light when it acts on a certain substrate. Burd introduced the luciferase gene into the mouse genome and placed it under the control of a DNA element, called the p16 promoter, that is primarily activated in senescent cells. Therefore, the senescent cells in these mice make luciferase. When the substrate for luciferase is injected, the mice produce light. Mice normally don't make photons, so the light emitted from the mice provides an estimate of senescent-cell abundance, X.

The experiment has several limitations, such as stronger absorption of light from inner regions, genetic disruption of the natural p16 system which enhanced the risk of cancer so the experiment could not probe very old ages, and experimental noise. But the experiment is a good starting point.

Looking at total light emitted from these mice as a measurement of senescent cells X, we see that X rises and falls around an increasing trajectory with age (Figure 7.3). This suggests two timescales: a fast timescale of fluctuations over weeks, and a slow timescale in which X rises over years (Figure 7.4). This fast-slow timescale separation, as mentioned above, is a key element in our model.

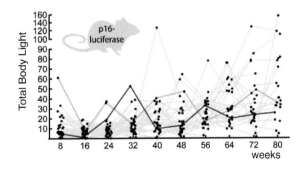

FIGURE 7.3 Dynamics of a senescent-cell reporter in 33 mice over 80 weeks. Light from a p16-luciferase reporter was measured every 8 weeks. Lines connect the data for each individual mouse. Green and purple lines are example trajectories. Adapted from Burd et al. (2013).

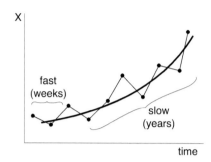

FIGURE 7.4   Senescent-cell data shows separation of timescales, with fast fluctuations over weeks around a slow rise over years.

Analyzing the data provides five quantitative patterns to test the theory:

i. The *average X grows at an accelerating rate* with age (Figure 7.5 shows the average of the data in Figure 7.3). Such accelerating accumulation of senescent cells with age is also seen in human tissues.

ii. The *variation of X between individuals rises with age* (Figure 7.6). Old mice have a larger range of X than young mice. Some old mice even have X levels similar to young mice. (Figure 7.6 shows the standard deviation of the data in Figure 7.3.)

iii. *Reducing relative heterogeneity:* the standard deviation of X grows more slowly than the mean. Thus, individuals become more similar to each other in relative terms. The standard deviation/mean, called the coefficient of variation (CV), drops with age (Figure 7.7); its inverse 1/CV rises approximately linearly with age (Figure 7.7 inset).

The declining heterogeneity means that damage X becomes *more similar in relative terms between individuals with age*. Such similarity might be expected for a core driver of aging. By contrast, downstream effects of X, like age-related diseases, depend on genetics and environment and are thus more variable between individuals with age.

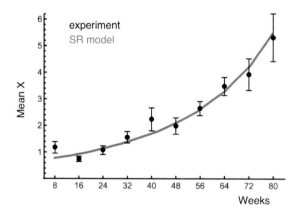

FIGURE 7.5   Mean level of senescent cells accelerate with age.

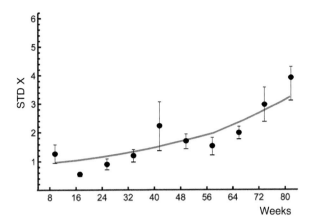

FIGURE 7.6   Standard deviation of senescent cells rises with age.

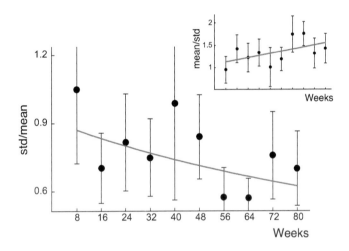

FIGURE 7.7   Coefficient of variation, defined as CV = std/mean, of senescent cells drops with age. Inset: 1/CV rises with age.

iv. Distributions of X between individuals at a given age are skewed to the right. There are more individuals above average X than below it (Figure 7.8). The skewness of these distributions gradually drops with age, as they become more symmetric.

v. The autocorrelation time of X increases with age (Figure 7.9 right). A mouse that is higher or lower than average stays so for longer periods the older it is (Figure 7.9 left panels). Thus, with age, the stochastic variation in X becomes more persistent.

Interestingly, these dynamical features are shared with the human frailty index described in the previous chapter. The mean and standard deviation of frailty rise with age, but relative heterogeneity drops with age, as does the skewness of the frailty distributions. In the next chapter, we will see that senescent-cell dynamics are tightly linked with the diseases and conditions that make up the frailty index.

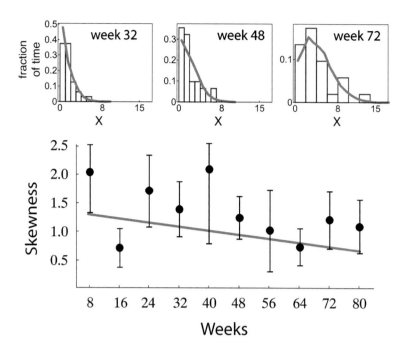

FIGURE 7.8   Damage distributions are skewed to the right and skewness drops with age.

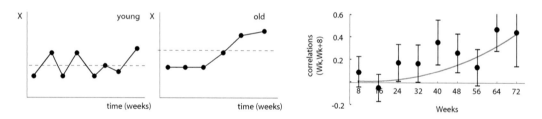

FIGURE 7.9   Damage fluctuates rapidly in young and becomes more persistent with age as seen in rising autocorrelation time.

THE SATURATING REMOVAL MODEL CAN EXPLAIN SENESCENT CELL DYNAMICS

These dynamical features of senescent cells can be explained by our simple truck model for aging, which we call the **saturating removal** model, developed by Omer Karin in his PhD with me (Karin et al. 2019). Omer scanned a wide class of models and found that the essential features that a model needs to explain the senescent-cell dynamics are precisely the truck model. In fact, these dynamical features are enough to constrain the possible models into a single minimal candidate.

The first feature is **two timescales**, one fast and one slow: X is produced and removed on a timescale that is much faster than the rise in DPU which occurs on the timescale of the lifespan. This separation of timescales allows us to write an equation for the rate of change of X in which the production and removal rates vary slowly and depend on age. The model also includes stochastic noise. Thus,

$$\frac{dX}{dt} = \text{production} - \text{removal} + \text{noise}$$

The production rate of X rises linearly with age in the model:

$$\text{production} = \eta\tau$$

where we use τ for age and t for time to make sure that we understand that there are two timescales: a fast scale (days-weeks) in which damage reaches steady state, and a slow timescale (years) over which production rate $\eta\tau$ changes.

The removal of X is carried out by removal processes, such as NK cells that kill senescent cells. If this removal worked at a constant rate β per senescent cell, the removal term would thus be $-\beta X$. However, this does not match the data. The equation is $\frac{dX}{dt} = \eta\tau - \beta X$, whose quasi-steady-state solution is a linear rise of X with age, $X_{st} = \eta\tau/\beta$. This is ruled out by the data which shows an accelerating rise with age (Figure 7.5).

Instead, the model explains the accelerating rise in X by a **removal rate that drops with the number of senescent cells**. In other words, removal saturates; senescent cells inhibit their own removal. Such a drop could be due to several processes: immune cells that remove senescent cells are downregulated if they kill too often, and they can be inhibited by factors that the senescent cells produce. The drop of removal rate can also be simply due to a saturation effect, in which the removing cells become increasingly outnumbered by senescent cells.

To model the saturation, we use a Michaelis–Menten form which is good both for inhibition due to secreted factors and for saturation (see Exercise 7.4)

$$\text{Removal} = \beta\frac{X}{\kappa + X}$$

where β is the maximal senescent-cell removal capacity, in units of senescent cells/time, and κ is the concentration of X at which they inhibit half of their own removal rate. The removal rate *per senescent cell* thus drops with senescent cells abundance, $\frac{\beta}{\kappa + X}$ (Figure 7.10), and the number of senescent cells lost per unit time is that rate times X, namely $\frac{\beta X}{\kappa + X}$.

Combining production and removal, we obtain a model for the rate of change of X:

(7.1)
$$\frac{dX}{dt} = \eta\tau - \beta\frac{X}{\kappa + X}$$

Note that this model assumes that maximal removal capacity β does not decline with age. Adding such a decline, namely $\beta(\tau)$, generally leaves the conclusions the same. For simplicity, we ignore this possibility and recall that NK cell numbers do not drop strongly with age in humans.

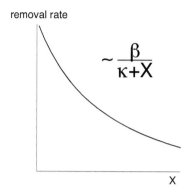

FIGURE 7.10   Removal rate per senescent cell drops with senescent-cell abundance X.

Let's compute the steady state of X. On the fast timescale of weeks, the production rate $\eta\tau$ can be considered as constant. Setting $dX/dt = 0$ in Eq. 7.1, we find that the (quasi-) steady-state X of is

$$(7.2) \qquad\qquad X_{st} \approx \frac{\kappa\eta\tau}{\beta - \eta\tau}$$

Thus, X_{st} rises linearly with age at first. Then, the term on the bottom becomes closer and closer to zero. The rise of X accelerates and diverges at a critical age $\tau_c = \beta/\eta$ (Figure 7.11). Figure 7.4 shows a good agreement between data in dots and the model in black.

When X levels rise high enough, they reach levels not compatible with life. Thus, the critical age $\tau_c = \beta/\eta$ where X diverges is a rough approximation for the mean lifespan. This equation indicates that lifespan can be extended by increasing the repair capacity β, or by reducing the senescent-cell production rate η. More trucks or slower construction of houses extend longevity.

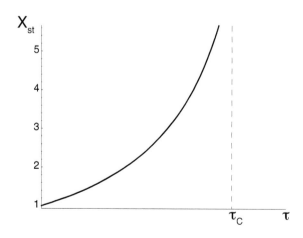

FIGURE 7.11   Mean senescent cells X accelerates and diverges at a critical age.

ADDING NOISE TO THE MODEL EXPLAINS THE VARIATION BETWEEN INDIVIDUALS IN SENESCENT-CELL LEVELS

If this model was all there was, then all individuals would age at the same rate and die at the same age. In other words, so far, the model does not explain why genetically identical organisms differ in the number of senescent cells. We need to describe the fluctuations of X over time which produce the differences between individuals. To understand these stochastic features of the dynamics, we need to consider the **noise** in the model.

The simplest way to model noise is to add a white-noise term ξ. This white-noise term adds and removes small amounts of X randomly, with mean zero and a variance described by the parameter 2ϵ. The factor 2 is for convenience in the equations below. The noise describes fluctuations in production and removal due to internal or external reasons such as injury, infection, and stress (cortisol). Variations in sleep, which affects repair, can also play a role. In fact, we don't currently know what the noise describes. White noise is a convenient way to wrap up our ignorance in a mathematical object that we can work with.

We arrive at the main model of this chapter and the next, a stochastic differential equation called the **saturating removal model**:

(7.3)
$$\frac{dx}{dt} = \eta\tau - \beta\frac{x}{x+\kappa} + \sqrt{2\epsilon}\,\xi$$

Let's take a nice deep sigh of relief to celebrate making it to this point.

THE SATURATING REMOVAL MODEL CAPTURES THE VARIATION BETWEEN INDIVIDUALS

We will now use this model to understand the dynamics of senescent cells, and then to understand the origin of the Gompertz law. Let's begin with the variation in X between individuals at a given age. To do so, we need to compute the damage distribution, $P(X)$.

There is an elegant way to compute it. Historically, this approach comes from analysis of particles undergoing Brownian motion under the influence of forces, known in physics as Langevin equations. I like this approach because it has a graphic interpretation in which X is the position of a particle rolling down a bowl-shaped contour (Figure 7.12).

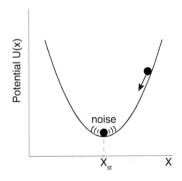

FIGURE 7.12 Damage dynamics can be likened to a particle in a bowl-shaped potential well jiggled by noise.

To describe the bowl, one recasts Eq. 7.3 using a potential function $U(X)$. The potential function provides the shape of the bowl. The slope of the potential is the restoring force that pushes the particle back to the equilibrium at the bottom of the bowl: the slope $-dU/dX$ is defined as production minus removal, $\dfrac{dU}{dX} = \eta\tau - \beta\dfrac{X}{X+\kappa}$. At the bottom of the bowl the slope is zero, and thus production equals removal, so that the bottom is the position of the steady state.

The required potential is $U(X) = (\beta - \eta\tau)X - \beta\kappa\log(\kappa + X)$. You can check this by taking the derivative $-dU/dX$ to verify that it gives $\eta\tau - \beta\dfrac{X}{X+\kappa}$.

The particle rolls down to the bottom of the bowl and noise randomizes its position (Figure 7.12). The distribution of X that we seek is then analogous to the Boltzmann law that applies to particles in a potential under thermal fluctuations. The noise amplitude ϵ plays the role of temperature, and we obtain $P(x) \sim e^{-U/\epsilon}$. The mathematical details are provided in Solved Exercise 7.1 at the end of the chapter.

At young ages, the senescent cells are like a particle in a steep potential well because removal is not saturated and pushes hard to reduce senescent cells. The particle is jiggled by noise in the narrow well, and so $P(X)$ is narrowly distributed around a low mean damage (Figure 7.13). With age, the potential well opens to the right, as production rate rises. The mean damage increases and so does its variance (Figure 7.13).

The saturating removal model thus predicts a distribution of damage that we can find by plugging in the potential into the Boltzmann term $P(X) \sim e^{-U/\epsilon}$ to obtain

(7.4)
$$P(X) \sim (\kappa + X)^{\beta\kappa/\epsilon}\, e^{-(\beta - \eta t)X/\epsilon}$$

This distribution rises with X, reaches a peak, and then falls exponentially. It is skewed to the right and matches the skewed distributions observed in the mouse data (Figure 7.8, upper panel, blue lines).

This distribution provides an improved estimate for the average X that takes noise into account,

(7.5)
$$\langle X \rangle \approx (\kappa\eta\tau + \epsilon)/(\beta - \eta\tau)$$

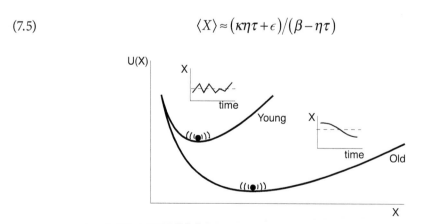

FIGURE 7.13   With age the bowl widens to the right, increasing damage mean and variance.

The mean X rises with age and diverges at age $\tau_c = \dfrac{\beta}{\eta}$. The standard deviation of X also rises with age and diverges at τ_c:

(7.6)
$$\sigma \approx \frac{\sqrt{\kappa\beta + \epsilon^2}}{\beta - \eta\tau}$$

This rise in the mean and standard deviation matches the observed rise with age of the standard-deviation in mice (Figures 7.5 and 7.6, blue lines).

Notably, the model even captures the fact that variation rises more slowly than the mean – the phenomenon of reducing relative heterogeneity. The ratio between the standard deviation and the mean of X, the coefficient of variation CV, drops with age as observed

$$CV = \frac{\sigma}{\langle X \rangle} \approx \frac{\sqrt{\kappa\beta + \epsilon^2}}{\kappa\eta\tau + \epsilon} \sim \frac{1}{\tau}.$$

This reducing heterogeneity is a distinguishing feature of the model. Most simple stochastic models show a constant CV.

The model also captures the increasing correlation times with age. At young ages, the bowl is steep. Thus, if X is away from steady-state X_{st}, it returns to it quickly (Figure 7.13). At old age, in contrast, the bowl is almost completely flat. The trajectory of the particle is dominated by noise, with little restoring force coming from the steepness of the bowl (Figure 7.13). Hence individuals that stray away from X_{st} have a weak restoring force and stay away for longer times.

Such increasing correlation times have a general name in physics, "**critical slowing down**.". They are a mark of an approaching phase transition. In the canonical example of a phase transition, the boiling of water, large and slow fluctuations in density can be seen near the boiling point. In other areas of science, slowing down of fluctuations can be a warning sign of a big transition. Examples include climate fluctuations before an ice age, or ecological fluctuations before a species extinction (Dai 2015). In our case, the phase transition is to death.

The mouse data allows estimating the four model parameters, η, β, κ, and ε. The best fit parameters are approximately $\eta = 4\ 10^{-4}$ days^{-2} $\sim 0.15\ /$ year $/$ day, $\beta = 0.3\ /$ day, $\kappa = 1$, $\epsilon = 0.1$, in units where the average senescent-cells level in young mice is 1. The rough estimate of lifespan $\tau_c = \dfrac{\beta}{\eta} \sim 2$ years is about right for mice.

These parameters give a specific prediction for the half-life of a senescent cell. The half-life is about 5 days in young mice and rises to about a month in old mice (25 days in 22-month old mice) because the trucks get overloaded. These predictions were intriguing enough to test experimentally, which Omer Karin did with senescent-cell expert Valery Krizhanovsky (Karin 2019). In young mice, the senescent-cell half-life was 5 ±1 days. In old mice (22-month-old), removal was much slower, with an estimated half-life of about a month. These measurements agree well with the predictions of the model.

GOMPERTZ MORTALITY WITH SLOWDOWN IS FOUND IN THE MODEL

We can now finally ask why do genetically identical organisms die at different times? What explains the exponential increase in risk of death with age (the Gompertz law), and the deceleration at very old ages (Figure 7.14)?

To connect senescent-cell dynamics to mortality, we need to know the relationship between senescent-cell abundance and the risk of death. The precise relationship is currently unknown. Let's therefore simply model death when senescent-cell abundance exceeds a threshold level X_C (Figure 7.15). The threshold represents a collapse of an organ system or a tipping point such as sepsis. Thus, death is a **first-passage time process**, when senescent cells cross X_C.

We use this threshold-crossing assumption to illustrate a way of thinking, because it provides analytically solvable results. Other dependencies between risk of death and senescent-cell abundance, such as Hill functions with various degrees of steepness, provide similar conclusions.

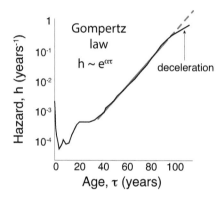

FIGURE 7.14   The risk of death rises exponentially according to the Gompertz law and decelerates at very old ages.

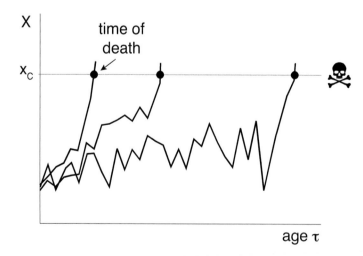

FIGURE 7.15   Death is modeled to occur when X crosses a threshold X_c.

Again, there is an elegant way to solve this first passage time problem and obtain the risk of death as a function of age. To estimate the probability that X crosses the death-threshold X_c, we apply an approach which is analogous to the rate of a chemical reaction crossing an energy barrier ΔG, namely the Boltzmann factor, $e^{-\frac{\Delta G}{k_b T}}$. In our case, the noise amplitude ε plays the role of temperature $k_b T$, and the energy barrier is the difference between the potential U at X_c and at the steady-state value X_{st}, $\Delta U = U(X_c) - U(X_{st})$. Thus, the probability per unit time for X crossing X_c, namely the risk of death that we call the hazard, is

$$h \approx e^{-\Delta U/\epsilon}$$

This equation is called Kramer's equation in the field of stochastic processes. An intuitive explanation is that the particle in the well needs to climb a potential difference of $\Delta U = U(X_c) - U(X_{st})$ to fall off into the death region (Figure 7.16). It needs to climb using "kicks" provided by the noise of size ϵ. Each noise kick can be either to the right or left. Since you need $\Delta U/\epsilon$ kicks, all in the right direction, the chance is exponentially small and goes as $e^{-\Delta U/\epsilon}$ (Figure 7.17).

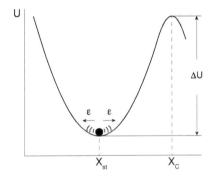

FIGURE 7.16   The particle needs to climb a potential difference of ΔU with noise kicks of size ε in order to fall off into the death region.

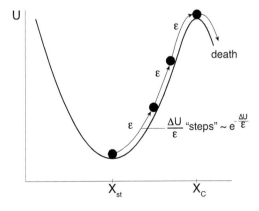

FIGURE 7.17   Death requires $\Delta U / \varepsilon$ kicks in the same direction, an event with exponentially small probability.

Now because the production rate of senescent cells rises linearly with age, it becomes easier and easier to reach the threshold X_c. As the potential bowl opens up, the "energy barrier" drops linearly with age, and risk therefore rises exponentially with age. The outcome for the risk of death (see Solved Exercise 7.1) is up to age-independent factors:

$$(7.7) \qquad\qquad h(\tau) \approx (\beta - \eta\tau)^{\frac{\kappa\beta}{\epsilon}+1} e^{\frac{(\kappa+X_C)\eta\tau}{\epsilon}}$$

This is a big moment. The hazard rises exponentially with age as $e^{\alpha\tau}$ with an exponent α, called the **Gompertz aging rate**, given by

$$\alpha = \frac{(\kappa + X_C)\eta}{\epsilon}.$$

We derived the Gompertz aging rate in terms of molecular parameters: noise, damage production rate and death thresholds. The aging rate is proportional to the slope of damage production with age, η. It is inversely proportional to the noise amplitude ϵ. Aging rate also rises with the death threshold X_c (see Solved Exercise 7.1 for further discussion).

The Gompertz law is thus intimately related to the linear rise with age of the production rate. If the production rate rose with age not linearly $\eta\tau$ but as age squared, for example, $\eta\tau^2$, we would not get the Gompertz law but instead a hazard that rises exponentially with age squared, $e^{\alpha\tau^2}$.

This solution also explains the observed deceleration of the hazard rate at very old ages (when $\eta\tau \approx \beta$). This deceleration is due to the prefactor $(\beta - \eta\tau)^{\frac{\kappa\beta}{\epsilon}+1}$ which becomes small as age approaches the critical age $\tau_c = \beta/\eta$. Fundamentally, slowdown occurs because near the critical age production and removal rates are nearly equal and cancel out, and noise becomes the primary driver of X. Therefore, damage dynamics resemble a random walk, with a nearly constant probability per unit time of crossing the threshold.[1]

The saturating removal model thus analytically reproduces the Gompertz law, including the deceleration at old ages (Figure 7.18). It gives a good fit to the observed mouse mortality curve using parameters that agree with the experimental half-life measurements and longitudinal senescent-cells data. The inferred threshold for death is $X_c = 17 \pm 2$, meaning that the threshold X_c is about 17 times larger than the mean senescent-cell level in young individuals.

In brief, genetically identical mice die at different times due to noise amplified by accumulating damage producing units and a saturating damage removal.

THE HUMAN GOMPERTZ LAW IS CAPTURED AS WELL

So far, we calibrated the model on mouse data. Let's now use it to study human mortality curves. We can ask which of the model parameters changes between mice and humans to provide the difference in their lifespans. A good description of human hazard curves, corrected for extrinsic mortality, is provided by the same parameters as in mice, except for a 60-fold slower increase in η, the slope of senescent-cell production with age. Figure 7.19 shows the model with added extrinsic mortality at a level indicated by the dashed line, in comparison to the observed mortality.

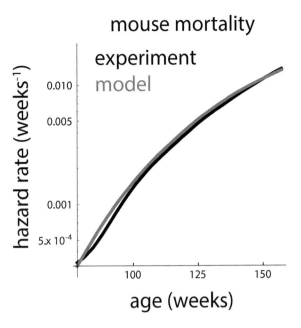

FIGURE 7.18   The saturating removal model analytically captures the Gompertz law and deceleration of mortality in mice.

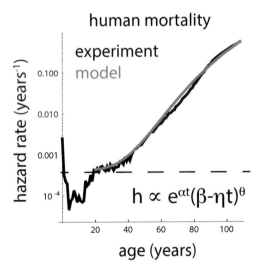

FIGURE 7.19   The saturating removal model with additive constant extrinsic mortality reproduces the Gompertz law and deceleration of mortality in humans.

The smaller value of η in humans can be due to improved DNA maintenance and enhanced damage repair in humans compared to mice. Thus, there are fewer DPUs at a given age, and each DPU gives rise to fewer senescent cells. Fewer houses each making less garbage.

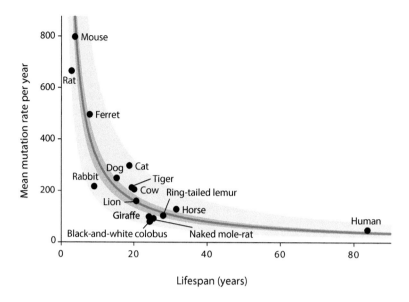

FIGURE 7.20 Mutation rate per stem cell is inversely proportional to lifespan in mammals. Adapted from Cagan et al. (2022).

Indeed, studies that compare mammalian species show that rates of DNA mutation, epigenetic alteration, and translation errors are all inverse to lifespan in different mammals (Figure 7.20) (Ke et al. 2017; Cagan et al. 2022; Ake Lu et al. 2021).

The theory therefore suggests that the parameter η is the main knob that evolution tunes in order to generate the lifespans of different mammals, as in the mass-longevity triangle of the previous chapter.

What practical use does the theory have? We end this chapter with ways in which the saturating removal model conceptualizes approaches to slow down aging. As we show in the next chapter, the model can address the use of drugs that eliminate senescent cells, known as senolytic drugs (Kirkland and Tchkonia 2020; Zhang et al. 2021). It provides the strength and frequency of treatment needed to reduce age-related diseases and addresses the question of whether starting treatment at old age can be effective. Stay tuned.

THE SATURATING REMOVAL MODEL EXPLAINS AGING PATTERNS ALSO IN ORGANISMS THAT LACK SENESCENT CELLS

The theory makes specific predictions, making it a falsifiable theory. If it fails to explain an experimental pattern, we throw it out. That is why we like to test the model against as many patterns as possible. To do so, we turn to a wider range of organisms.

It is remarkable that diverse organisms show the Gompertz law and the other patterns of aging we saw in the previous chapter. Even single-celled organisms show the Gompertz law under certain conditions. Examples include yeast cells with asymmetric replication called budding (Steinkraus, Kaeberlein, and Kennedy 2008).

These model organisms lack senescent cells but still show the nearly universal features of aging shared with mice and humans. To understand this, recall that the model is agnostic about the molecular nature of the damage X and the damage-producing units. It can therefore be generalized beyond senescent cells. It should apply to any form of damage that is causal for aging as long as it has saturating-removal-type dynamics, namely production rate that rises linearly with age and saturating removal.

We therefore compare the model to key experiments in organisms without senescent cells such as yeast, the fruit fly *Drosophila melanogaster* and the roundworm *C. elegans*.

Let's think of how the saturating removal model can apply to single-celled organisms or simple animals. We need a DPU, a factor that accumulates in cells with time and is irremovable, and which generates damage X. Surveying the massive experimental research on model organisms shows recurring themes which can serve as candidate DPUs. One of these candidates is aggregates of misfolded proteins which the cell is unable to remove. These aggregates grow inside the cell and cause damage such as mitochondrial damage and reactive oxygen species (Labbadia and Morimoto 2015).

Let's use protein aggregates as an example. The mass of the aggregates, $P(t)$, grows when misfolded proteins join the aggregate. Since the cell makes proteins at a constant rate, and each protein has a chance to be misfolded due to transcription/translation errors, P grows at a constant rate, $dP/dt = a$, and thus $P = at$. Indeed, older single-celled organisms tend to have larger aggregates.

The aggregates P are toxic to cells. They produce damage X, whatever its exact nature, at rate b per unit aggregate. Damage production is thus $bP = abt$. We can therefore define the parameter η, the slope of the rise in production rate with as, as $\eta = ab$. According to law 2, the removal mechanisms of the damage X saturate, and we have the saturating removal model, $dX/dt = \eta\, t - \beta\, X/(\kappa + X) + \text{noise}$.

Other molecular factors can play the role of the DPUs, such as epigenetic alterations in yeast. We conclude that the model can be plausible also for single-celled model organisms and invertebrates that lack senescent cells.

RAPID SHIFTS BETWEEN HAZARD CURVES

As we saw in the previous chapter, shifting fruit flies to a life-extending diet led to a rapid switch to the better survival curve within a couple of days. The sins of the past are forgiven (Figure 6.12). In contrast, shifting temperature only changed the slope of the death curve, and the accumulated past hazard was not forgotten (Figure 6.13).

These phenomena can be explained by the saturating removal model based on the rapid turnover of damage X and the irremovable nature of damage producing units.[2] The plummet to a lower curve after a shift of diet in Figure 7.21A can be explained if the diet lowers the amount of damage produced by each damage-producing unit, P. The life-extending diet slows the rate of living and shifts the tradeoff from growth to repair, so that each P makes less damage X per unit time, namely a lower value of the parameter b. It's as if the denizens of the houses begin to conserve, reuse, and recycle and thus produce less garbage.

Therefore, after the shift, the production term $\eta t = aP(t)$ is lower, and the hazard curve shifts down. The time it takes to move between curves is the rapid turnover time of the damage X, on the order of hours to days in flies. When recycling begins, the trucks clear the excess garbage rapidly. The model predicts that this transient time should grow with age due to slowdown of removal.

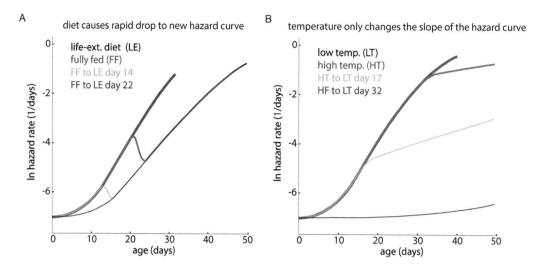

FIGURE 7.21 Saturating removal model captures the rapid shift to the better hazard curve when diet is changed to a lifespan-extending diet, and a change of hazard slope upon a shift to lower growth temperature. (A-B) described in text

In contrast, the slope change of hazard induced by temperature in Figure 7.21B can be explained by a reduced rate of production of new damage producing units – slower construction of new houses. This is a reduction in the accumulation parameter a in the equation $dP/dt = a$, without a change in b, the rate of damage production per unit P. Thus, P keeps its accumulated number but is produced more slowly starting from the temperature shift at time t_0. Thus, $P(t) = a\, t_0 + a'\,(t - t_0)$, where a' is the reduced production rate. This results in the hazard curve shown in Figure 7.21B.

SCALING OF SURVIVAL CURVES

A further test is whether the model can explain the **scaling** of survival curves for C. elegans, yeast, and mice under different life-extending or life-shortening genetic, environmental, and diet perturbations. Recall that scaling means that the survival curves collapse on the same curve when age is scaled by mean lifespan. Few things get physicists more excited than a good data collapse, where different curves fall on top of each other when normalized. Such a collapse reveals that perturbations that are superficially unrelated to each other actually impact the system in the same way.

Most interventions that increase lifespan show scaling. For example, caloric restriction shows survival curves that collapse nearly onto the same curve when age is normalized by median lifespan (Figure 7.22). Scaling seems to apply in all organisms tested for caloric restriction.

The model provides this scaling property to an excellent approximation, for perturbations that affect the production rate parameter η (Figure 7.23A). Lowering η lengthens lifespan but preserves the shape of the survival curve.

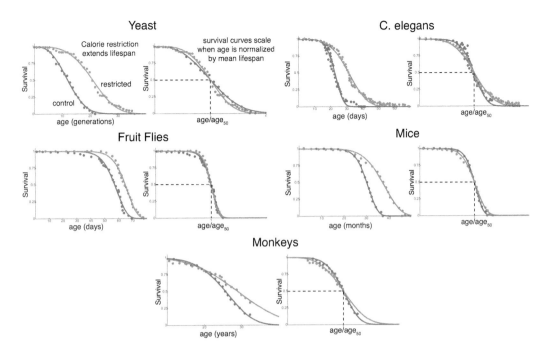

FIGURE 7.22 Caloric restriction increases lifespan with scaling of survival curves in different species. Adapted from Conn's *Handbook of Models for Human Aging* (Second Edition; Chapter 19).

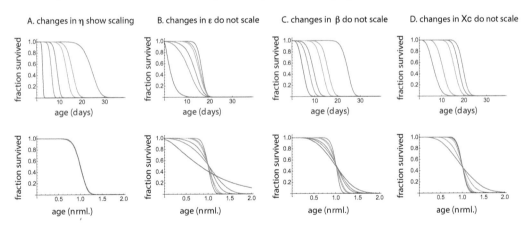

FIGURE 7.23 Scaling is found in the saturating removal model upon changes in damage production slope, but not changes in other parameters, the removal rate β, noise ε...or death threshold X_c.

Caloric restriction is indeed thought to reduce the rate of damage production by slowing down metabolic activity, and shifting the cells into increased repair and recycling, as mentioned above. This should reduce the rate of DPU production and toxicity per DPU, reducing η.

The pathways for the effects of caloric restriction include the IGF1 pathway mentioned in Chapter 6. Mutations in this pathway, such as daf-2 mutations in *C. elegans*, were among the first life-span-extending mutations found, as you can read in the history by pioneer aging researcher Cynthia Kenyon (2011). These mutations shift the entire survival curve to

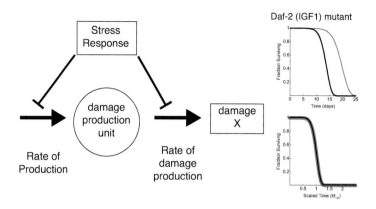

FIGURE 7.24   Inhibiting the IGF1 pathway affects production of damage, increasing lifespan with survival-curve scaling.

longer lifetimes but maintain its shape showing scaling (Figure 7.24). Similar longevity and scaling are seen in IGF1 mutations across organisms.

Interestingly, there is no scaling in the model when a perturbation affects other parameters, such as removal rate β (Figure 7.23B). Thus, there is no scaling when the trucks are affected. When removal rate β is increased – more trucks – the model predicts that lifespan increases and the survival curve becomes steeper. The reason for the steeper curve is that damage has a shorter half-life due to faster removal, and thus there is less time for noise to randomize things. Survival becomes more deterministic, and organisms die at more similar times.

Since scaling is so common in yeast, worms, and mice, we may conclude that many perturbations affect houses – and thus damage production rate, η – but much fewer affect the trucks. Finding mutations or interventions that do not scale can help to pinpoint the identity of the trucks and garbage in each organism.

One prediction is that removing senescent cells by senolytic drugs should make mice not only live longer but make their survival curves steeper. This is because senolytic drugs increase the effective removal rate β of senescent cells. Steeper survival curves are indeed found in senolytic treatment as shown in Figure 7.25. The steeper curve can be seen by noting that the treatment increases median lifespan more than maximal lifespan. The model passes this test as well.

Another intervention in mice that increases lifespan and steepens the survival curve is inhibiting the age-associated loss of blood vessels. Inducing growth of blood vessels by mild overexpression of the vascular regulator VEGF (Grunewald et al. 2021) caused a nearly 40% increase in mouse lifespan (Figure 7.26), with reduced senescent-cell load, delayed cancer, and improved organ function. Increased blood vessels have many positive effects including better oxygen flow to organs. One of the effects of increased blood vessels is more roads for the trucks – access for the immune cells that remove senescent cells, thus potentially increasing their removal rate β and explaining the steeper survival curves.

Other treatments that protect vasculature also enhance lifespan and steepen the survival curve in mice, especially in males. This includes treatments that reduce glucose spikes that damage blood vessels, such as a ketogenic diet and the diabetes drug acarbose.

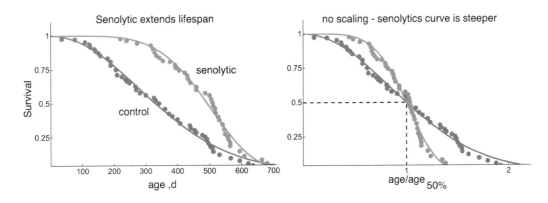

FIGURE 7.25   Senolytic drugs increase mean mice lifespan and show steepened survival curves as predicted by model. Adapted from Kowald and Kirkwood (2021).

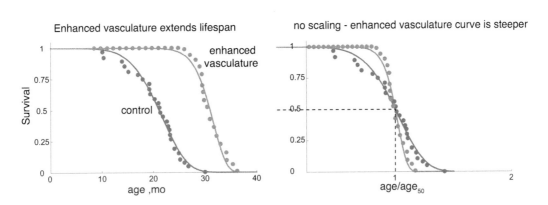

FIGURE 7.26   Enhanced vasculature in mice increases mean lifespan and shows steepened survival curves. Adapted from Grunewald et al. (2021).

APPROACHES TO SLOW DOWN AGING AND AGING-RELATED DISEASES

Current medicine treats age-related diseases one at a time: diabetes, cancer, heart disease, and so on. A different approach would be to deal with their largest risk factor – to slow the aging process itself, or more precisely to slow the rise of senescent cells and other aging-related damage. This is the **Geroscience hypothesis**: *slowing the core process of aging will prevent and mitigate multiple age-related diseases in one fell swoop.*

The conceptual framework we discussed points to two general strategies: reduce production rate η or increase removal capacity β.

Reducing production rate – the houses – can be achieved by boosting cellular damage-repair systems (Figure 7.27). One way to achieve this is caloric restriction and other types of restricted feeding. As mentioned above, fasting shifts the balance from growth toward maintenance in cells. Effort is devoted to developing drugs that mimic caloric restriction by, for example, perturbing the IGF1 pathway; these drugs include mTOR inhibitors. One such drug is metformin, used for treating diabetes since the 1920s. Metformin inhibits the

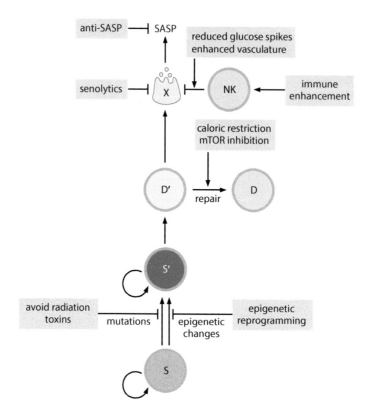

FIGURE 7.27 Approaches to slow down aging target damage production and removal processes.

IGF1 pathway and tips the balance toward more repair. An encouraging sign is that people taking metformin have lower risk of cancer. A current effort is to convince the federal food and drug administration (FDA) to allow clinical trials for aging; currently only trials for specific diseases are allowed. Metformin is one suggested drug for such a trial, along with other mTOR inhibitors such as rapamycin.

Increasing the removal rate of senescent cells is also an attractive possibility. That is what **senolytic drugs** do. Senolytics remove senescent cells by exploiting the Achilles heels of senescent cells that are not found in most other cells. One such drug, for example, inhibits an anti-cell-death pathway called Bcl2, exploiting the fact that this pathway helps senescent cells resist death to a greater extent than most other cells in the body. Senolytics are entering clinical trials in humans for specific diseases (Zhang et al. 2022a).

The number of garbage trucks and hence the maximal removal capacity β can also be increased by immune-based strategies. Recently, an immune approach used to fight cancer cells was repurposed to remove senescent cells – enhancing the trucks – in mice (Amor et al. 2020). In this approach, called CAR-T, killer T cells are taken from the mouse and genetically engineered to express a receptor that recognizes a protein found only on the surface of senescent cells. These engineered T cells were reintroduced into the mice and killed senescent cells. Another potential approach targets the factors that senescent cells secrete, such as pro-inflammatory factors, and reduces their effect.

It would be elegant to reduce production at its root, the DPUs, by targeting the altered stem cells, but this is challenging. Approaches that mitigate epigenetic changes, such as cell reprogramming or drugs that modulate enzymes that add or remove epigenetic marks, have shown promise in certain aging contexts (Eisenstein 2022). Some of the reprogramming approaches currently face unwanted side effects, however, because they revert cells to primordial stem-like states that may increase the risk of cancer.

Non-pharmacological approaches that offer a degree of rejuvenation have been known for ages. These are the quartet of exercise, healthy diet, good sleep and reduced stress by means of social relationships, purposeful action, psychotherapy, meditation, and moderation. Exercise has coordinated beneficial effects such as lowering insulin resistance and reducing excess fat in tissues. Healthy diet likewise reduces fat and insulin spikes. Easing the mind reduces stress including the activity of the HPA axis and sympathetic nervous system. This positively impacts insulin resistance and blood pressure and enhances the maintenance roles of the immune system.

The road to extended healthspan is long. As mentioned above, there is likely a series of causal factors, each with a different type of houses, garbage and trucks, and addressing one factor reveals the next like peeling an onion. There is probably a brain-specific factor related to Alzheimer's disease. Damaged neurons accumulate with age and set off neuroinflammation, raising the specter of prevalent dementia at ages above 90 as a price of increased longevity due to amelioration of cancer and heart disease.

Furthermore, with age, organs like the skin, gut, and lung are increasingly composed of small local "kingdoms" of cells, each from a different clone of stem cells. Each clone arises from an altered stem cell that has a growth advantage relative to its neighbors. Each kingdom thus has its own individual random mutations. In people above 70, for example, most blood cells are made from a few stem-cell clones that took over the bone marrow (Mitchell et al. 2022). Blood health depends on the luck of which mutations these stem cells happen to have. Thus, although senescent cells are a major component, other factors are likely to be important for aging.

In this intense chapter, we described a theory of aging. It explains why genetically identical organisms die at different times. The theory provides a first-principle understanding of universal patterns of aging such the Gompertz law and scaling of survival curves. The biological features of the theory have wide generality: linearly rising damage production and saturating removal. It can thus apply to senescent cells in humans and also to other forms of damage in invertebrates without senescent cells. The theory explains how different perturbations alter survival curves and conceptualizes ways to extend healthspan.

Speaking of healthspan, in the next chapter we explore age-related diseases.

EXERCISES

Solved Exercise 7.1: Compute the distribution of damage X at a given age in the saturating removal model

The distribution of X, denoted $P(X)$, is the probability of having X senescent cells. To calculate $P(X)$, we use a method that applies to any stochastic differential equation of the form:
$\dfrac{dX}{dt} = v(x) + \sqrt{2\epsilon}\,\xi$. In the saturating removal model, the "velocity" $v(x)$ equals production

minus removal, namely $v(X) = \eta\tau - \beta X / (\kappa + X)$. The idea is to rewrite the equation using a **potential** $U(X)$, whose slope is determined by the velocity as follows: $\dfrac{dU}{dX} = -v(X)$.

The potential function has the shape of a bowl (Figure 7.14). The variable X is like a ball rolling in the bowl (Figure 7.14). The ball rolls down the slope, with velocity $-v(x)$ given by the slope of the bowl, dU/dX. The steeper the bowl, the faster the ball rolls. The bowl can be considered to be coated with a thick goo, as described in Steven Strogatz's superb book on dynamical systems (Strogatz 2001) and so the ball slogs through the goo in a damped way and settles down at the minimum of the bowl without oscillating. At the minimum point the slope is zero, $dU/dX = 0$, and that is where the steady state is, $X = X_{st}$. The steeper the sides of bowl, the faster the ball returns to X_{st} when it is perturbed.

Let's now add noise. Noise jiggles the ball position X so that it deviates from X_{st}. These jiggles cause a distribution of X values, $P(X)$. Again, the steeper the bowl, the less noise can move X away from X_{st}, and the narrower the distribution $P(X)$.

The nice thing about the potential-function way of writing the equation is that we can easily compute the steady-state distribution. This distribution $P(X)$ for a given age τ is given by the Boltzmann distribution, with ϵ playing the role of temperature:

(7.8)
$$P \propto e^{-U(X)/\epsilon}$$

An intuitive explanation is provided in Solved Exercise 7.3. The shallower the bowl, or the larger the "temperature," the wider the distribution $P(X)$.

For the saturating removal model, the potential $U(X)$ is

(7.9)
$$U(X) = (\beta - \eta\tau)X - \beta\kappa \log(\kappa + X)$$

We can safely assume that age τ is constant over the fast timescale needed to reach the steady-state distribution $P(X)$, except at very old ages. Plotting $U(X)$ shows that at young ages the bowl is steep, and therefore, the distribution is localized around the mean (Figure 7.28). With age, the bowl becomes more and more shallow, because its right-hand slope drops as $-\eta\tau$. At the critical age, when $\eta\tau = \beta$, the bowl opens up and the mean X at steady state goes to infinity.

Plugging $U(X)$ from Eq. 7.9 into the Boltzmann-like law of Eq. 7.8, we obtain the distribution

(7.10)
$$P(X) \propto e^{-\frac{(\beta - \eta\tau)X}{\epsilon}} (\kappa + X)^{\frac{\beta\kappa}{\epsilon}}$$

which reaches a peak and then falls exponentially with X. This distribution of senescent cells in the model is skewed to the right, consistent with the skewed distributions observed in the mouse data (Figure 7.10, blue lines). With age, the skewness of this distribution drops (Figure 7.11 blue line).

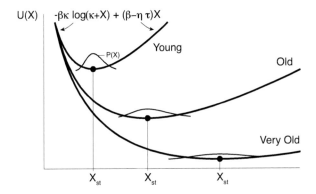

FIGURE 7.28   The effective potential in the saturating removal model widens to the right with age.

Exercise 7.2: Show that the saturating removal model gives the Gompertz law of mortality

To estimate the probability that X crosses the death-threshold X_c, we apply Kramers' equation

$$h \approx e^{-\frac{U(X_c)-U(X_{st})}{\epsilon}}$$

The potential U in our model is given by Eq. 7.9. For the Gompertz law to hold, one needs the term $\frac{U(X_c)-U(X_{st})}{\epsilon}$ to decrease linearly with age τ, so that $h \approx e^{\alpha\tau}$.

The exponent of the hazard rate in the model indeed shows the required linearity in time. This is the factor $\eta\tau$ in this complicated expression:

$$(7.11) \qquad -\frac{U(X_C)-U(X_{st})}{\epsilon} = \frac{(\kappa+X_C)\eta\tau - X_C\beta + \kappa\beta\log\left(\frac{(\kappa+X_C)(\beta-\eta\tau)}{\kappa\beta}\right)}{\epsilon}$$

which can be written, up to a prefactor that does not depend on age, as:

$$(7.12) \qquad h(\tau) \approx e^{-\frac{\beta X_C}{\epsilon}} (\beta-\eta\tau)^{\frac{\kappa\beta}{\epsilon}+1} e^{\frac{(\kappa+X_C)\eta\tau}{\epsilon}}$$

The prefactor $e^{-\frac{\beta X_C}{\epsilon}}$ indicates that the hazard at age zero drops exponentially with the threshold X_c. Thus, although X_c increases the logarithmic slope of the hazard $\alpha = \frac{(\kappa+X_C)\eta\tau}{\epsilon}$, the intercept at age zero drops exponentially with X_c. As a result, the hazard at a given age is reduced overall by increasing the threshold X_c. This makes sense since a higher threshold makes it more difficult for X to cross X_c.

Exercise 7.3: Intuitive derivation of the Boltzmann-like form of the steady-state distribution

Consider a stochastic process of the form $\frac{dX}{dt} = v(x) + \sqrt{2\epsilon}\xi$. The function $v(x)$ is called the velocity of x. Define the potential $U(x)$ by $\frac{dU}{dx} = -v(x)$. Explain intuitively why, at steady state, the probability distribution is $P(x) = P_0 \, exp\left(-\frac{U(x)}{\epsilon}\right)$.

Solution: Consider a large number of particles moving along a one-dimensional pipe. They diffuse with diffusion coefficient ϵ and are also swept along the pipe by a velocity field $v(x)$. The particle density at steady state is $P(x)$. The flux at point x due to the velocity field is the velocity times the density: $v(x)P(x)$. The flux due to diffusion can be found by Fick's law of diffusion, which shows a diffusive flux from high to low densities proportional to the gradient: $-\epsilon dP/dx$. At steady state, total flux is zero, so that the two fluxes must sum to zero: $v(x)P - \epsilon dP/dx = 0$. Thus, $\frac{dP}{dx} = \frac{v(x)P(x)}{\epsilon}$. The solution is $P(x) = P_0 exp\left(-\frac{U(x)}{\epsilon}\right)$

Thus, at steady state, in regions where velocity is large, the density $P(x)$ shows a steep opposing slope so that diffusion flux – the effects of noise – can balance velocity flux.

Exercise 7.4: Removal of senescent cells based on saturating their own removal process

Senescent cells are removed by immune cells such as NK cells, which we will denote by R. There are a total of R_T removing cells in the body, and that this number does not change appreciably with age. The R cells meet senescent cells, denoted X, at rate k_{on} to from a complex $[R\,X]$ which can either fall apart at rate k_{off} or end up killing senescent cells at rate v. Thus, $R + X \rightleftarrows [RX] \rightarrow R$.

a. Explain the following dynamic equation for the complex:

$$\frac{d[RX]}{dt} = k_{on} R \ X - \left(v + k_{off}\right)[RX]$$

b. Use the fact that R cells can be either free or in a complex, so that $R + [RX] = R_T$, to show that the removal rate of senescent cells is

$$Removal = \frac{\beta X}{k + X}$$

c. Compute the maximal removal capacity β, and the half-way saturation point k. Explain intuitively.

Exercise 7.5: No repair

Consider an accumulation process of damage with constant production and no removal

$$\frac{dX}{dt} = \eta + \sqrt{2\epsilon}\xi$$

a. What is the mean damage X as a function of age?

b. What is the distribution $P(X)$?

c. What is the hazard assuming that death occurs when $X > X_c$? Is there a Gompertz law?

Exercise 7.6: Age-dependent reduction in repair capacity

Consider a process in which damage is produced at a constant rate η, and removal does not saturate. Removal rate per cell drops with age,

$$\frac{dX}{dt} = \eta + (\beta - \beta_1 \tau)X + \sqrt{2\epsilon}\xi.$$

a. What is the mean damage X?

b. What is the distribution $P(X)$ at age?

c. What is the ratio of mean and standard deviation of X: $<X>/\sigma$?

d. What is the hazard, if death occurs when $X > Xc$? Is there a Gompertz law?

Exercise 7.7: Deterministic model

Assume that the Gompertz law arises not from stochastic effects, but instead from individual differences, set at birth, in X production and removal parameters, in which each individual i has its own noise-free equation $\frac{dX}{dt} = \eta_i - \beta_i X$. Death is modeled to occur when X crosses threshold X_c. What distribution of production and removal parameters η_i, β_i. can provide the Gompertz law? What patterns of aging does this model not explain?

Exercise 7.8: Parameter effects

What is the effect on the hazard curve of the saturating removal model of a change in each of the parameters $\beta, \eta, \epsilon, \kappa$? Plot examples of hazard curves to demonstrate your answer.

Exercise 7.9: Senescent-cell half-life

Show that in the saturating removal model, the half-life of a senescent cell is

$$t_{1/2} = log\,(2)(\kappa\beta + \epsilon)/\beta(\beta - \epsilon\tau)$$

Exercise 7.10: Critical slowing down

Read (Scheffer et al. 2009).

a. How does critical slowing down relate to the model?

b. Suggest a phenomenon beyond those discussed in Scheffer which might show critical slowing down and suggest an experiment or measurement to test this.

Exercise 7.11: (Challenging question) General model

Damage is produced at rate $\eta(X, \tau)$ and removed at rate $\beta(X, \tau)$. The equation is

$$\frac{dX}{dt} = \eta(X, \tau) + \beta(X, \tau) + \sqrt{2\epsilon}\xi$$

a. What is the steady-state distribution at age τ?

b. What is the risk of death as a function of age, modeled by first passage time of a threshold X_c?

c. Under which conditions does risk of death go as the Gompertz law?

Exercise 7.12: Strehler and Mildvan (1960) model for the Gompertz law

Strehler and Mildvan (1960) (SM) proposed a phenomenological process for the Gompertz law. Organisms are assumed to start with an initial survival capacity, termed the vitality V, that declines linearly with age τ as $V(\tau) = V_0(1 - B\tau)$, where B indicates the fraction of vitality loss per unit time. Animals experience random external challenges or insults with a mean frequency K. Challenges have random magnitudes, exponentially distributed with an average magnitude D that expresses the average deleteriousness of the environment. Death occurs when the magnitude of a challenge exceeds the remaining vitality. A review of the SM theory can be found in (Finkelstein 2012).

a. Show that these assumptions produce the Gompertz law $h(\tau) = ae^{b\tau}$. Calculate a and b.

b. What similarities and differences does this theory have with the saturated removal model?

Exercise 7.13: Heterochronic parabiosis

Parabiosis is the surgical joining of two mice so that they share circulation. When joining a young and an old mouse, known as heterochronic parabiosis, the young mouse shows signs of aging whereas the old mouse rejuvenates. Senescent cell abundance decreases in the old mouse, whereas it increases in the young mouse (Karin and Alon 2021).

a. Explain these effects using the model. Hint: the trucks from the young mouse can help the old mouse.

b. The survival curve of the old mouse is made steeper and that of the young mouse less steep compared to a control experiment joining two equal-aged mice. Explain using the model.

c. Develop an extended model that describes heterochronic parabiosis. Assume that senescent cells stay in the tissues and are not shared by the mice, but that the immune cells that remove them are shared through the circulation. Plot senescent cells in each of the two mice as a function of time after joining.

Exercise 7.14: Aging rates of different organs
Based on the model, would you expect that different organs in the same individual will age at similar rates or different rates? Explain (100 words).

Exercise 7.15: Numerical solution and decreasing twilight

a. Write a computer program to solve the SR model.

b. Suppose that death occurs when X first crosses X_c, and illness occurs when X first crosses X_d, with $X_d < X_c$. In 100 runs of the simulation, record the time of illness and time of death.

c. Does average remaining lifespan after illness, known as twilight, increase with age of illness onset or decrease? Answer: twilight should decrease with the age of illness.

d. Explain the decreasing twilight phenomenon in the model (50 words).

This exercise requires numerically simulating a stochastic ordinary differential equation. The main idea is to use Euler's method (Chapter 1, Exercise 1.6), with timestep dt, and to add a noise term consisting of a random number times a timestep $\sqrt{dt}$. The square root is because noise acts like a random walk or diffusion process, which moves a mean distance of $\sim\sqrt{dt}$ in a time interval dt.

Here is a simple algorithm to get you started. Suppose you wish to solve the equations $\frac{dx}{dt} = f(x) + \sqrt{2\epsilon}\,\xi$ with initial condition $x(0) = 0$.

$x(0) = 0$ (initial condition)
$t(0) = 0$ (time$=0$)
$N = 1000$ (total time steps)
$dt = 0.1$ (0.1s intervals)
for $i = 1{:}N$
    $t(i) = t(i-1) + dt$
    $r(i) = $ random_number
    $dx(i) = f(x(i-1))\,dt + \sqrt{2\epsilon}\,r(i)\,\sqrt{dt}$
    $x(i) = x(i-1) + dx(i)$
end

NOTES

1 Note that the analytical approximation begins to be inaccurate when ηt approaches β, and simulations of the full model are needed to compute the hazard curve at old ages. Simulations show that the rise of the hazard curve slows and converges to a constant hazard at very old ages.

2 A parameter set for flies is $\beta = 1$ /hour, $\epsilon = 1$ /hour and $\eta = 0.03$/hour day, $X_c = 15$. Flies on a life-extending diet have a lower production slope $\eta = 0.02$ /hour day.

FURTHER READING

Karin et al. 2019. "Senescent cell accumulation slows with age providing and explanation for the Gompertz law."

REFERENCES

Ake Lu, Viviana Perez, Zhe Fei, Ken Raj, and Steve Horvath. 2021. "Universal DNA Methylation Age Across Mammalian Tissues." *Innovation in Aging* 5 (Supplement_1): 410. https://doi.org/10.1093/geroni/igab046.1588.

Amor, Corina, Judith Feucht, Josef Leibold, Yu-Jui Ho, Changyu Zhu, Direna Alonso-Curbelo, Jorge Mansilla-Soto, et al. 2020. "Senolytic CAR T Cells Reverse Senescence-Associated Pathologies." *Nature* 583 (7814): 127–32. https://doi.org/10.1038/s41586-020-2403-9.

Brauning, Ashley, Michael Rae, Gina Zhu, Elena Fulton, Tesfahun Dessale Admasu, Alexandra Stolzing, and Amit Sharma. 2022. "Aging of the Immune System: Focus on Natural Killer Cells Phenotype and Functions." *Cells* 11 (6): 1017. https://doi.org/10.3390/cells11061017.

Burd, Christin E., Jessica A. Sorrentino, Kelly S. Clark, David B. Darr, Janakiraman Krishnamurthy, Allison M. Deal, Nabeel Bardeesy, Diego H. Castrillon, David H. Beach, and Norman E. Sharpless. 2013. "Monitoring Tumorigenesis and Senescence in Vivo with a P16 INK4a-Luciferase Model." *Cell* 152 (1–2): 340–51. https://doi.org/10.1016/j.cell.2012.12.010.

Cagan, Alex, Adrian Baez-Ortega, Natalia Brzozowska, Federico Abascal, Tim H. H. Coorens, Mathijs A. Sanders, Andrew R. J. Lawson, et al. 2022. "Somatic Mutation Rates Scale with Lifespan across Mammals." *Nature* 604 (7906): 517–24. https://doi.org/10.1038/s41586-022-04618-z.

Eisenstein, Michael. 2022. "Rejuvenation by Controlled Reprogramming Is the Latest Gambit in Anti-Aging." *Nature Biotechnology* 40 (2): 144–46. https://doi.org/10.1038/d41587-022-00002-4.

Finkelstein, Maxim. 2012. "Discussing the Strehler-Mildvan Model of Mortality." *Demographic Research* 26:191–206. https://doi.org/10.4054/DemRes.2012.26.9.

Grunewald, M., S. Kumar, H. Sharife, E. Volinsky, A. Gileles-Hillel, T. Licht, A. Permyakova, et al. 2021. "Counteracting Age-Related VEGF Signaling Insufficiency Promotes Healthy Aging and Extends Life Span." *Science* 373 (6554): eabc8479. https://doi.org/10.1126/science.abc8479.

Karin, Omer, and Uri Alon. 2021. "Senescent Cell Accumulation Mechanisms Inferred from Parabiosis." *GeroScience* 43 (1): 329–41. https://doi.org/10.1007/s11357-020-00286-x.

Karin, Omer, Amit Agrawal, Ziv Porat, Valery Krizhanovsky, and Uri Alon. 2019. "Senescent Cell Turnover Slows with Age Providing an Explanation for the Gompertz Law." *Nature Communications* 10 (1): 5495. https://doi.org/10.1038/s41467-019-13192-4.

Ke, Z., P. Mallik, A.B. Johnson, F. Luna, E. Nevo, D. Zhang, V.N. Gladyshev, A. Seluanov, and V. Gorbunova. 2017. "Translation Fidelity Coevolves with Longevity." *Aging Cell*, 16(5): 988–993.

Kenyon, Cynthia. 2011. "The First Long-Lived Mutants: Discovery of the Insulin/IGF-1 Pathway for Ageing." *Philosophical Transactions of the Royal Society B: Biological Sciences* 366 (1561): 9–16. https://doi.org/10.1098/rstb.2010.0276.

Kirkland, J. L., and T. Tchkonia. 2020. "Senolytic Drugs: From Discovery to Translation." *Journal of Internal Medicine* 288(5): 518–536.

Kowald, Axel, and Thomas B. L. Kirkwood. 2021. "Senolytics and the Compression of Late-Life Mortality." *Experimental Gerontology* 155 (November): 111588. https://doi.org/10.1016/J.EXGER.2021.111588.

Labbadia, Johnathan, and Richard I. Morimoto. 2015. "The Biology of Proteostasis in Aging and Disease." *Annual Review of Biochemistry* 84:435–64. https://doi.org/10.1146/annurev-biochem-060614-033955.

Mitchell, Emily, Michael Spencer Chapman, Nicholas Williams, Kevin J. Dawson, Nicole Mende, Emily F. Calderbank, Hyunchul Jung, et al. 2022. "Clonal Dynamics of Haematopoiesis across the Human Lifespan." *Nature* 606 (7913): 343–50. https://doi.org/10.1038/s41586-022-04786-y.

Scheffer, Marten, Jordi Bascompte, William A. Brock, Victor Brovkin, Stephen R. Carpenter, Vasilis Dakos, Hermann Held, Egbert H. Van Nes, Max Rietkerk, and George Sugihara. 2009. "Early-Warning Signals for Critical Transitions." *Nature* 461:53–9. https://doi.org/10.1038/nature08227.

Steinkraus, K. A., M. Kaeberlein, and B. K. Kennedy. 2008. "Replicative Aging in Yeast." *Annual Review of Cell and Developmental Biology* 24:29–54. https://doi.org/10.1146/annurev.cellbio.23.090506.123509.

Strehler, B. L., and A. S. Mildvan. 1960. "General Theory of Mortality and Aging." *Science* 132 (3418): 14–21. https://doi.org/10.1126/science.132.3418.14.

Strogatz, Stephen. 2001. *Nonlinear Dynamics and Chaos: With Applications to Physics, Biology, Chemistry, and Engineering, Second Edition (Studies in Nonlinearity)*. Vol. 32. Westview Press. https://doi.org/10.5860/choice.32-0994.

Unnikrishnan, Archana, Sathyaseelan S. Deepa, Heather R. Herd, and Arlan Richardson. 2018. "Chapter 19 – Extension of Life Span in Laboratory Mice." In *Conn's Handbook of Models for Human Aging (Second Edition)*, edited by Jeffrey L. Ram and P. Michael Conn, 245–70. Academic Press. https://doi.org/10.1016/B978-0-12-811353-0.00019-1.

Yang, Yifan, Ana L. Santos, Luping Xu, Chantal Lotton, François Taddei, and Ariel B. Lindner. 2019. "Temporal Scaling of Aging as an Adaptive Strategy of *Escherichia Coli*." *Science Advances* 5. https://doi.org/10.1126/sciadv.aaw2069.

Zhang, Lei, Louise E. Pitcher, Vaishali Prahalad, Laura J. Niedernhofer, and Paul D. Robbins. 2022a. "Targeting Cellular Senescence with Senotherapeutics: Senolytics and Senomorphics." *FEBS Journal* 290 (5): 1362–83. https://doi.org/10.1111/FEBS.16350.

Zhang, Lei, Louise E. Pitcher, V. Prahalad, Laura J. Niedernhofer, and Paul D. Robbins. 2021. "Recent Advances in the Discovery of Senolytics." *Mechanisms of Ageing and Development* 200: 111587.

CHAPTER 8

Age-Related Diseases

MANY DISEASES OCCUR ALMOST exclusively at old age. These **age-related diseases** include cancer, osteoarthritis, failure of specific organs such as heart failure, kidney failure, and lung failure, and neuro-degenerative diseases such as Alzheimer's disease and Parkinson's disease.

In this chapter, we will understand why age is the major risk factor for these diseases and explore the universality of their dynamics. We will also discuss treatment. These diseases are currently treated one by one, and we will discuss how future medicine can take a major step forward by treating aging itself in order to address all of these diseases at once.

Age-related diseases are diverse and affect different systems. It is therefore striking that they share a common pattern in their incidence curves. Incidence is the probability to get the disease at a given age. It is calculated by considering 100,000 people without the disease at age *t* and asking how many will be diagnosed over the following year.

And now for the pattern shared between hundreds of age-related diseases:

The incidence of age-related diseases rises exponentially with age and drops at very old ages (Figure 8.1). The slope of exponential increase is similar for different diseases, but not identical, around 3%–8% per year.

Understanding this exponential rise is a major aim of this chapter. We need to understand why age 20 is different from age 70 in ways that make these diseases so much more likely. We will also understand why incidence drops at very old ages.

Another goal of this chapter is to explain the causes of several diseases of unknown origin. In doing so we will see mathematical analogies between diseases. This will form columns in the periodic table of diseases featured in the next chapter of our book.

DISEASES CAUSED BY THRESHOLD CROSSING OF SENESCENT CELLS HAVE AN EXPONENTIAL INCIDENCE CURVE

To understand age-related disease incidence, we will use a simple model based on the senescent cell theory of the previous chapter. This model was developed by Itay Katzir during his PhD with me (Katzir et al. 2021).

DOI: 10.1201/9781003356929-14

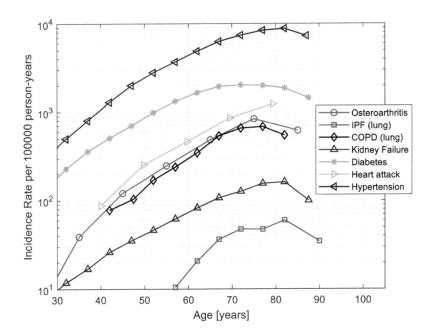

FIGURE 8.1   The incidence of age-related diseases rises exponentially with age and drops at very old ages.

The basic idea is that **diseases of old age are due to a phase transition in which aging pushes a parameter of a physiological circuit across a threshold.** Once the threshold is crossed, the circuit behavior changes dramatically: cells grow without control as in cancer or die without control as in degenerative diseases.

Aging indeed affects the parameters of physiological circuits. In particular, senescent cell load X induces systemic inflammation and reduces regeneration, which changes circuit parameters. Above a certain level of senescent cells, the circuit undergoes a bifurcation-its steady state becomes unstable and pathology emerges. Therefore, in the model, a disease occurs when senescent cells cross a threshold that is specific for each disease. We call this threshold the **disease threshold** X_d.

Although each disease has its own threshold X_d, the underlying senescent cell dynamics are common to all diseases. These dynamics are described by the saturating removal model of Chapter 7, Eq. 7.3. When the concentration of senescent cells X crosses the disease threshold, the individual gets the disease (Figure 8.2). Each individual crosses the threshold at different times, due to the stochastic nature of the dynamics of senescent cells.

The time of disease onset is therefore the time when senescent cell concentration first crosses the threshold X_d – a first-passage time problem.

Conveniently, in the previous chapter we already solved this first-passage-time problem. The solution is an exponential hazard curve – the Gompertz law – that slows at very old ages. The probability of crossing the threshold X_d rises exponentially with age, with an exponential slope of approximately

$$\alpha \approx \eta X_d / \epsilon$$

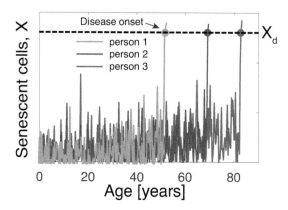

FIGURE 8.2   In the disease-threshold model, a disease occurs when senescent cells cross a disease-specific threshold.

where η and ϵ are the senescent cell production and noise parameters.

This explains the exponential rise of disease incidence curves. Since diseases have different exponential slopes, each disease has its own threshold X_d. The disease threshold must not exceed $X_{death} = 17$, otherwise the model would predict that death precedes the disease, and we would not observe the disease.

DECLINE OF INCIDENCE AT VERY OLD AGES IS DUE TO POPULATION HETEROGENEITY

If this were all, everyone would cross the disease threshold in the model and get the disease. In reality only a fraction of people ever do. This is where the second parameter in the model comes into play – only a fraction ϕ of the population are **susceptible**. The parameter ϕ ranges between zero and one. Some conditions are rare with low ϕ, others like hypertension and osteoarthritis are common, with ϕ exceeding 0.1. The precise value of the susceptibility depends on genetic and environmental factors, as we will discuss.

The susceptible fraction stems from the notion of population heterogeneity in the fields of epidemiology and genetics. People differ in their risk for a given disease. To model this, we assume that only a fraction ϕ of the population has a low disease threshold X_d. The remaining population has higher values of the disease threshold that are not reached during normal aging. We call these the *non-susceptible* fraction of the population.

The susceptible fraction explains the decline of incidence curves at very old ages. Recall that incidence is computed from the population without the disease. At very old ages, *most of those that are susceptible have already had the disease*. This results in the decline in incidence rate.

The model thus has two parameters for each disease: the disease threshold and the susceptibility. Let's solve the model for the incidence curve to see where the rise and fall originate (see Solved Exercise 8.1 for more details about the approximations involved). The idea is that incidence $I(t)$ is approximately equal to the hazard $h(t)$ – the probability to cross the disease threshold X_d at age t, multiplied by the disease-free survival curve $F(t)$ – the fraction of the population who still did not get the disease. Thus $I(t) = h(t) F(t)$. Since h rises

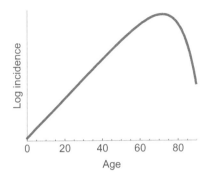

FIGURE 8.3   In the disease threshold model, incidence rises exponentially and drops at very old ages.

and F declines, their product is a curve with a maximum. Writing disease-free survival in terms of hazard results in an equation for the incidence

$$I(t) = \phi \, h(t) e^{-\int_0^t h(t)dt}$$

and by plugging in a Gompertz-like hazard $h = h(0)e^{\alpha t}$, we obtain an analytical incidence formula

(8.1)
$$I = \phi h(0) \, e^{\alpha t} e^{-\frac{h(0)}{\alpha}\left(e^{\alpha t}-1\right)}$$

At first, incidence rises exponentially (Figure 8.3), until at very old ages the last term dominates, since it is an exponential of an exponential, and incidence plummets.

Note that susceptibility ϕ simply multiplies the incidence in Eq. 8.1 and thus determines its overall height; the shape of the incidence curve, including its slope, intercept, and age of peak incidence, is determined by a single parameter – the disease threshold X_d. Using the saturated removal model of the previous chapter, we can find how the disease threshold determines the shape parameters in Eq. 8.1: to a good approximation, the slope is $\alpha = 0.009 \, X_d - 0.02$ and the hazard intercept is $log_{10} \, h(0) = 4.14 - X_d$ for the relevant range of disease thresholds X_d between 10 and 16 (Katzir et al. 2021).

Armed with Eq. 8.1, we can now find the best-fit values of X_d and ϕ for a given empirical incidence curve and see how well the disease-threshold model captures the data.

THE MODEL DESCRIBES WELL THE INCIDENCE CURVES OF AGE-RELATED DISEASES

To test this model requires a global set of incidence curves. We turn to the large medical-record database from Clalit health services that we used in Chapter 3 on hormone seasonality. The data includes about 900 disease categories, each found in the records of at least 10,000 people. The categories are international disease codes, called ICD9 level 2. Of these, about 200 diseases rise at least 20-fold between ages 30 and 80, and can be defined as strongly age-related diseases.

These diseases include some of the most common age-related conditions such as Parkinson's disease, glaucoma, congestive heart failure, end-stage renal disease, liver cirrhosis, cataract, hypertension, and osteoarthritis (Figure 8.4).

The disease-threshold model captures the data well (Figure 8.4). It captures more than 90% of the variation in over 90% of these diseases. The goodness of fit has a median of $R^2 = 0.97$, where $R^2 = 1$ is a perfect fit. The typical disease threshold values X_d range between 12 and 16.

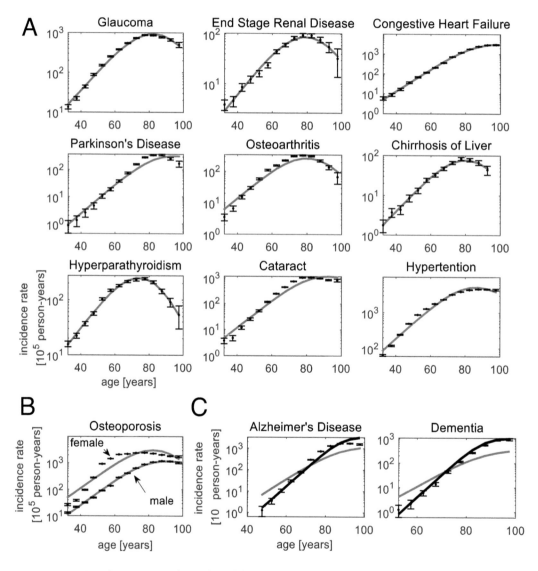

FIGURE 8.4   Incidence curves from the Clalit health record dataset are well-described by the disease threshold model (A). Exceptions are (B) female osteoporosis which rises after menopause and (C) Alzheimer's disease and dementia which require a threshold higher than the death threshold (black line). Adapted from Katzir et al. (2021).

The model does not, however, describe well the incidence of several age-related diseases. A notable example is osteoporosis in women (Figure 8.4B). The incidence curve rises sharply after age 50, due to effects related to menopause, in a way that the model cannot capture. On the other hand, osteoporosis in men is well described by the model (Figure 8.4C). This suggests that menopause-related changes go beyond the current framework.

An interesting case occurs in Alzheimer's disease and dementia. The incidence curves of these diseases have an exceptionally large slope of about 20% per year. The model can only explain this large slope with a disease threshold $X_d = 20$ that exceeds the threshold for death $X_{death} = 17$ (black line in Figure 8.4C). The best fit with the maximal X_d values equal to the death threshold underestimates the incidence slope (blue lines in Figure 8.4C).

This suggests that the age-related factor X in dementia might be distinct from total body senescent cells, and has its own saturating-removal dynamics. This makes sense because the brain is a protected organ with its own version of immune function. One candidate for this brain-specific damage might be accumulation of prion-like protein aggregates in neurons which saturate their mitochondrial-based removal systems. This is consistent with the damaged mitochondria and protein aggregates that are found in neurodegenerative diseases.

All in all, the model explains an astonishingly large fraction of the incidence curves of age-related diseases.

To understand what the disease-threshold model is capturing, let's explore in more detail the patterns in the incidence data. One such pattern concerns the timing of the peak incidence, and its relationship to the slope of the incidence curve. Naively, one may think that the steeper the slope, the earlier the peak incidence – steeper curves max out earlier (Figure 8.5). But the data shows otherwise: the steeper the curve, the later the peak incidence. Why? Because steeper incidence curves begin lower, as defined by their intercept, namely the extrapolated incidence at age zero (See Figure 8.5, age = 0).

Remarkably, the disease-threshold model exhibits this pattern. The steeper the slope, as described by a higher disease threshold X_d, the later the peak incidence (Figure 8.6). The reason is that the slope rises linearly with the disease threshold, but the intercept at age

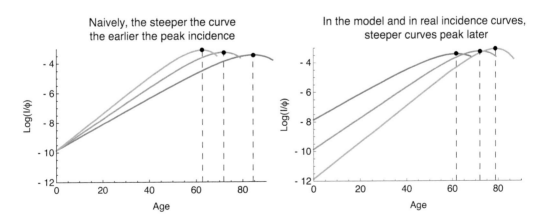

FIGURE 8.5 Naively, steeper incidence curves should peak earlier. In the data and the model, steeper curves peak later.

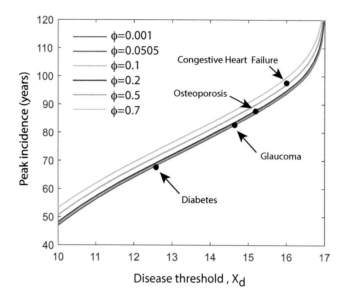

FIGURE 8.6 Age of peak incidence rises with disease threshold in the model and data.

zero $I(0)$ drops exponentially with this threshold. To understand this, recall the analogy with a particle in a potential well: a high threshold makes it exponentially harder for noise to generate enough senescent cells to cross the threshold at young ages; the zero intercept thus decays exponentially with threshold, namely $I(0) \sim e^{-\beta X_d/\epsilon}$.

Thus, the disease-threshold model captures some of the deep patterns in the data with only two free parameters per disease, of which only one, X_d, affects the shape of the curve. This is impressive.

But how does each specific disease occur when senescent cells cross a threshold? We need to link senescent cells and the physiology of each disease. To do so, we now focus on several classes of pathologies and specify, for each case, the mechanism for their onset at the threshold-crossing.

We begin with cancer and infection. We then consider an age-related disease in which the lungs fail, called Idiopathic Pulmonary Fibrosis (IPF). Its cause is a mystery. We will use our approach to explain this disease as an outcome of fundamental principles of tissue homeostasis. We will then show that a seemingly unrelated disease of the joints, osteoarthritis, belongs to the same "mathematical class" as IPF.

CANCER INCIDENCE CURVES CAN BE EXPLAINED BY THRESHOLD CROSSING OF TUMOR GROWTH AND REMOVAL RATES

Cancer risk rises by 4000% between age 25 and 65. The incidence curves of most cancer types show the familiar exponential rise with age and drop at very old ages. To explain this in our model, we need to find out why cancer is like a threshold-crossing phenomenon, and how senescent cells can push physiology across this threshold.

Cancer cells arise continuously in the body due to accumulation of mutations. If conditions are right, the mutant cells grow faster than their neighbors. These cancer cells are removed by immune surveillance, primarily by the innate immune cells such as NK cells and macrophages, and at later stages by adaptive immunity including *T* cells. If the cancer cells manage to grow beyond a critical number of roughly 10^6 cells, they organize a local microenvironment that can prevent further immune clearance.

A classic explanation for the age-dependence of cancer is called the **multiple-hit hypothesis**: the need for several mutations in the same cell to turn it into a cancer cell (Armitage and Doll 1954; Nordling 1953). Most cancers require a series of mutations, called oncogenic mutations, in order to knock-out pathways that prevent the cell from growing out of control. Such a multiple-hit process has a likelihood that rises roughly as the age to the power of the number of mutations. Cancer in the young often occurs because one of the mutations is already present in the germline and thus in all cells of the body.

This multiple hit hypothesis, however, cannot explain why incidence drops at very old ages. It also fails to explain why cancers which require a single mutation, such as some leukemias, also have an exponentially rising incidence with age. Even colon cancer, the poster child for a multiple-mutation progression, has exponentially rising incidence with age rather than a power law.

The present theory can provide a mechanism for the incidence curves of cancers. Consider cancer cells that proliferate at rate *p* and are removed at rate *r* (Figure 8.7). The rate of change of the number of cancer cells *C* equals proliferation minus removal:

$$\frac{dC}{dt} = pC - rC$$

Cancer grows when proliferation exceeds removal, $p > r$, and shrinks otherwise (Figure 8.8). This is just the knife-edge equation we saw in Chapter 2.

Both growth and removal of cancer are affected by senescent cell load *X*. With age, rising senescent cell levels inhibit the capacity of the immune system to remove cancer cells. The main cells that remove early cancer cells, NK cells and macrophages also remove senescent cells. They become saturated when senescent cells become abundant and cannot keep up with the demand for cancer removal services. The garbage trucks are overloaded. Thus, removal rate *r* drops with the number of senescent cells, $r = r(X)$.

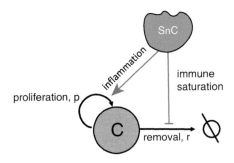

FIGURE 8.7   Cancer cells proliferate and are removed. Senescent cells increase proliferation and decrease removal.

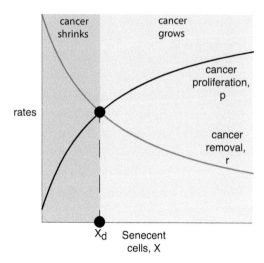

FIGURE 8.8   Senescent cells reduce removal and increase proliferation. When they exceed a threshold, cancer starts growing exponentially.

A second cancer-inducing effect is chronic inflammation caused by the factors that senescent cells secrete. One may think of many cancers as an AND-gate between chronic inflammation and oncogenic mutations. Inflammation reduces the growth rate of healthy cells, giving mutant cancer cells a relative growth advantage. Many cancers arise only after chronic inflammation causes cells to become less differentiated – to undergo metaplasia. Thus, inflammation can raise cancer proliferation rate p, so that proliferation rises with senescent cell levels $p=p(X)$.

Both effects, increasing proliferation p and lowering removal r, push cancer toward the threshold where proliferation exceeds removal. The senescent cell level where this occurs is our disease threshold X_d (Figure 8.8).

Individuals susceptible to a given form of cancer include those with genetic factors (e.g., BRCA mutations for breast and ovarian cancer) and exposure to environmental factors such as smoking for lung cancer and UV for skin cancer. These factors increase the probability of sporadic occurrences of the cancer cells in the tissue. The proliferation rate, p, and removal rate, r, both depend on conditions in the local tissue niche, as well as the mutational and epigenetic state of the cell. Hence, the more occurrences of cancer cells in the tissue, the higher the chance that $p>r$ for one of these cells, allowing it to proliferate and generate a tumor.

Cancer incidence is well documented, allowing a good test for theory. One comprehensive database, called SiteSEER, has incidence curves of 100 cancer types in the US. Of these cancers, 87 are at least mildly age-related. Of these, 66 are well-described by the disease threshold model ($R^2 > 0.9$) (Figure 8.9). The typical values of X_c are 13–15, and the susceptibilities for different types of cancer range from 10^{-4} to 0.1.

There are several types of cancer with a poor fit to the model (Figure 8.9b), namely cancers that are common at young ages such as testicular cancer, Hodgkin's lymphoma, and cervical cancer (which has a viral origin).

All in all, the disease-threshold model seems to describe a wide range of age-related cancers very well.

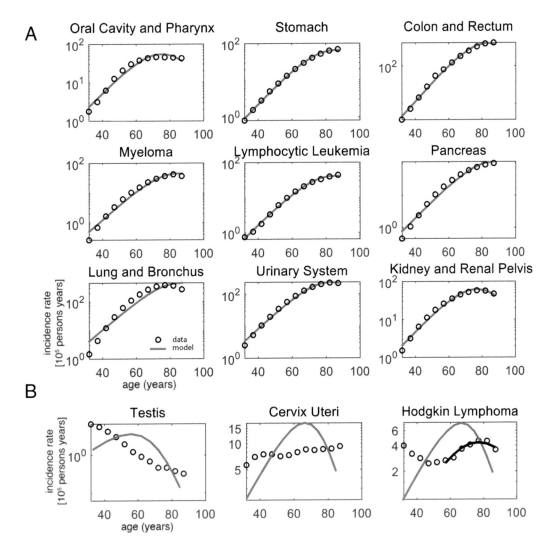

FIGURE 8.9 The disease-threshold model captures the incidence curves of most cancer types (A). The exceptions (B) are cancers with early age of onset. Adapted from Katzir et al. (2021).

MANY INFECTIOUS DISEASES HAVE AGE-RELATED MORTALITY

A general theory such as the disease-threshold model can be used to make connections between very different diseases. To demonstrate this connection across disease classes, we consider infectious diseases, such as pneumonia, flu, and COVID-19. In many infectious diseases, mortality rate rises exponentially with age (Figure 8.10).

Infections are diverse. Each pathogen has ingenious ways to resist the immune system. But despite this complexity, pathogens share a mathematical unity, which is analogous to the cancer model we just saw.

A virus or bacterium has proliferation rate, p, because all pathogens come from pathogens. It is removed at rate r by the immune system. The number of pathogens N thus obeys the same knife-edge equation as cancer cells, $dN/dt = (p-r)N$.

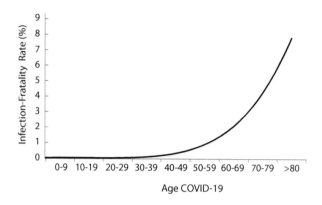

FIGURE 8.10   Mortality from many infections rises exponentially with age, as shown for COVID-19.

Infections become deadly when they grow exponentially, that is when $p>r$. The host is killed by damage caused directly by the pathogen, or more commonly by the collateral damage unleashed by the immune system trying to fight the pathogen.

In young individuals, pathogen removal usually exceeds proliferation. The pathogen is handily eliminated by the immune system. However, just as in the case of cancer, senescent cells X can reduce the removal rate $r(X)$ in multiple ways. Senescent cells overload the immune cells, including NK cells and macrophages, whose job is to fight pathogens. They also contribute to the decline of the adaptive immune system, including T cells, with age.

Such effects lower the removal rate of the pathogen, so that $r(X)$ decreases with X. At old age, a critical threshold X_d is reached, where removal equals proliferation $r(X_d)=p$. Beyond this threshold a given infection that would be removed at young ages now has $p>r$ and grows exponentially.

Thus, the age-dependence of both cancer and infection belongs to the same mathematical class – they are eliminated at young ages but have a phase transition to growth at a critical point X_d, giving rise to the observed incidence curves.

Let's now turn to a different class of diseases, progressive fibrotic diseases. But first, to recognize that we are doing a lot of work here, let's take a nice deep sigh of relief.

A THEORY FOR IPF, A DISEASE OF UNKNOWN ORIGIN

A striking feature of the disease-threshold theory is that it can offer new explanations for age-related diseases that are poorly understood. To see this, we consider IPF, which stands for **idiopathic pulmonary fibrosis.** Its very name indicates that the cause is unclear: "Idiopathic" means disease of unknown cause, "pulmonary" means lungs, and "fibrosis" means excess scarring.

In IPF, lung capacity is progressively lost due to the scarring of tissue that is essential for breathing (Martinez et al. 2017). It is a chronic progressive disease that has no cure; patients often die within 1–3 years. The lifetime susceptibility to IPF is about $\phi=10^{-4}$. Its incidence rises exponentially with age and then drops (Figure 8.1).

To understand IPF, let's survey the relevant organ structure. The lung is made of branching tubes that end in small air sacs called **alveoli** (Figure 8.11). The alveoli let oxygen from the air go into the blood and let CO_2 out. The alveoli are made of a thin epithelial layer that

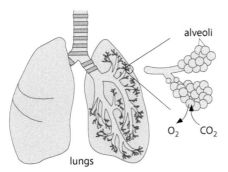

FIGURE 8.11 The lung is made of branching tubes called bronchi that end in alveoli that perform gas exchange.

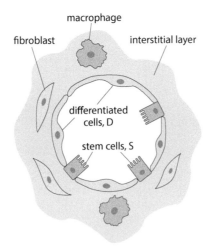

FIGURE 8.12 Alveoli have a thin epithelial layer surrounded by an elastic interstitial layer.

is one-cell thick surrounded by an interstitial layer. IPF scarring occurs in the interstitial layer around the alveoli (Figure 8.12).

The thin epithelial layer is made of two types of cells. The first cell type (alveolar type-1 cells) are large flat barrier cells, which we will call the differentiated cells D. The second type (alveolar type-2 cells) are smaller stem-like cells we will call S (Figure 8.13). These stem cells can divide to form new S cells or differentiate into D cells. The S cells also secrete a soapy surfactant that shields the cells from air particles and prevents collapse of the alveoli when we exhale.

The interstitial layer around the alveoli contains fibroblasts and macrophages, the stars of Chapter 5 on fibrosis. Macrophages are ready to gobble up bacteria and particles that make it through the layer of S and D cells. The fibroblasts produce the fibers which make the elastic sheath around the alveoli.

When there is injury to the D cells, they signal (with molecules such as TGF-beta) to S cells coaxing them to differentiate into new D cells (Figure 8.14). These injury signals also cause S cells to activate inflammation in the interstitial layer to start a healing process.

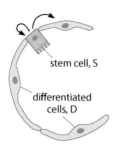

FIGURE 8.13   The epithelial layer is made of stem cells that renew themselves and produce differentiated cells.

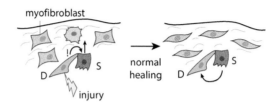

FIGURE 8.14   An injury causes alveolar cells to induce stem-cell division and signal fibroblasts to become scar-forming myofibroblasts. In normal healing, myofibroblasts decay and the tissue is restored.

The S cells signal the fibroblasts to become activated myofibroblasts, which proliferate and secrete extra fibers.

In normal healing, once the new D cells are made, the excess fibroblasts undergo programmed cell death, and the extra fibers are removed. S cells divide and renew the tissue, and the injury is repaired.

In IPF, an unknown factor causes an ongoing injury. The S cells multiply and reach higher numbers relative to D cells than in normal alveoli (Figure 8.15). They activate the fibroblasts to multiply and lay down excessive fibers, causing fibrosis. The interstitial tissue around the alveoli becomes a thick scar that reduces the ability of oxygen and CO_2 to flow in and out. It makes the alveoli stiff and less able to expand and contract. Eventually more and more alveoli become dysfunctional, leading to lung failure.

A major unknown in IPF is the origin of the injury. We can use what we have learned so far to make a theory for the source of the injury and explain why the risk of IPF rises exponentially with age, and why it occurs in only a small fraction of the population. We

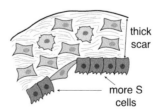

FIGURE 8.15   In IPF, stem cells proliferate and hyper-activate myofibroblasts causing excessive scarring and loss of alveolar function.

rely on research that shows that senescent cells are important for IPF: the affected alveoli have enhanced cellular senescence, especially in S cells (Martinez et al. 2017), and removing senescent cells by senolytic drugs reduces fibrosis in IPF mouse models (Hernandez-Gonzalez et al. 2021; Lopes-Paciencia et al. 2019).

We will thus explore how the accumulation of senescent cells might cause IPF. The main idea is that senescent cells slow down the rate of stem-cell proliferation; when stem-cell proliferation rate drops below removal rate, both S and D cell populations vanish – the alveolar tissue locally reaches zero cells.

STEM CELLS MUST SELF-RENEW AND SUPPLY DIFFERENTIATED CELLS

To understand IPF, we thus need to understand how stem-cell-based tissues work. Stem cells are found in organs that need to generate large numbers of cells. One class of such organs are barrier organs exposed to the outside world, like the lung, intestine, and skin. Because of this exposure, cells can be damaged and need to be replaced.

These organs divide labor: the majority of cells, D, do the main tissue work, and the minority (1%–5%) are stem cells, S, in charge of regenerating the D cells and themselves. Thus $S \rightarrow D$.

Stem-cell-based tissues differ from the organs we considered in part 1 of the book, where differentiated cells like adrenal cortex cells gave rise to their own kind, without need for stem cells (Figure 8.16).

Recall that in such tissues steady state requires that cell proliferation rate equals cell removal rate, otherwise the tissue grows or shrinks. In contrast, in stem-cell based tissues, the proliferation of stem cells S must *exceed* their removal, because some of the S divisions are needed to make the D cells. For stem cells, therefore, proliferation must balance two processes: stem-cell removal plus differentiation (Figure 8.16).

The stem-cell removal rate in many tissues is low because the stem cells are in a **protected niche**, where they are shielded from damage. Examples include the blood stem cells hidden in the bone marrow, the skin stem cells in the deep epithelium, and the gut stem cells tucked away at the bottom of crypts (Figure 8.17).

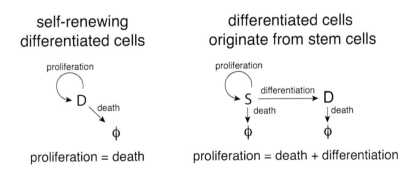

FIGURE 8.16 Comparison of circuits for self-renewing cell types and cell types renewed by stem cells.

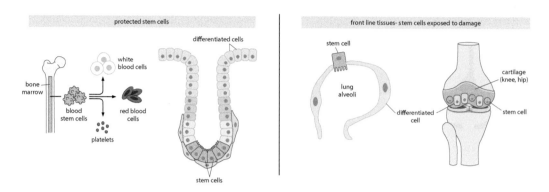

FIGURE 8.17 Comparison of organs with protected stem cells and organs with front-line stem cells exposed to damage.

In contrast, the lung alveoli are an example of a tissue where stem cells are on the **front lines**. Stem cells and differentiated cells are both exposed to damage, such as air particles, pathogens, and the mechanical stress of breathing. There is no other choice: the alveoli must be a thin monolayer of cells to allow diffusion of gasses and can't afford a deep layer for the stem cells. We call such tissues "**front-line tissues.**"

We are now ready to propose a mechanism for IPF.

INCIDENCE OF IDIOPATHIC PULMONARY FIBROSIS CAN BE EXPLAINED BY STEM-CELL REMOVAL EXCEEDING PROLIFERATION

In front-line tissues, stem cells are exposed to damage and removed often. Homeostasis is harder to achieve than in tissues in which stem cells are protected, because of the high rate of removal of stem cells.

To understand this, let's analyze the circuit that maintains organ size in front-line tissues. We will see that front-line tissues crash when removal exceeds proliferation.

Let's write down the basic equations (Figure 8.18) (Katzir et al. 2021). These equations account for stem-cell proliferation at rate p, and their differentiation to make differentiated cells D at rate q. The removal rate of S and D cells is r:

(8.2)
$$\frac{dS}{dt} = pS - rS - qS$$

(8.3)
$$\frac{dD}{dt} = qS - rD$$

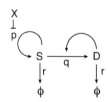

FIGURE 8.18 Circuit for front-line tissue, in which stem cells and differentiated cells are both removed, and differentiated cells feedback on the stem cells to maintain homeostasis. Senescent cell accumulation reduces stem-cell renewal.

Note that differentiation means that an S cell is lost and a D cell is gained. As a result, the $-qS$ term in the first equation, namely the rate of differentiation of an S to a D cell, shows up as a $+qS$ term in the second equation.

To maintain the proper amounts of S and D cells, there is a feedback loop. As mentioned above, D cells signal to S cells by secreting factors like TGF-beta that increase the rate of differentiation q, and thus $q=q(D)$. This feedback acts to restore homeostasis when cell numbers are perturbed, as analyzed in Solved Exercise 8.2. Pioneering work on such stem-cell circuits is due to Arthur Lander and colleagues (Lander et al. 2009).

We will now see that this circuit has a failure point. It breaks down when proliferation p falls below removal r – the cell population shrinks exponentially. To see this mathematically, we bound our equation from above by a simpler equation which declines to zero. We first add the two equations Eqs. 8.1 and 8.2 to get an equation for the total number of cells $S+D$

$$\frac{d(S+D)}{dt} = pS - rS - r\,D = pS - r(S+D)$$

This addition eliminates the feedback term $q(D)$, so our conclusions will work for any form of feedback! We increase the right-hand-side by changing S to $S+D$ because $S+D$ is always greater than S,

$$\frac{d(S+D)}{dt} < p(S+D) - r(S+D) = (p-r)(S+D)$$

We end up with the knife-edge equation for total number of cells T=S+D

$$\frac{dT}{dt} = (p-r)T.$$

Thus, when the proliferation rate falls below removal, $p<r$, the total cell number is bounded below an equation that goes to zero exponentially fast with time. Both S and D must go to zero (Figure 8.19).

After the collapse, tissue repair cannot proceed by regeneration because there are no more stem cells. Instead the tissue resorts to processes such as fibrosis, cell migration, and metaplasia, which are doomed to fail. Fibrosis reduces tissue function and pathology occurs.

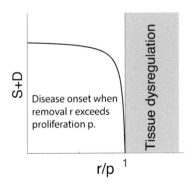

FIGURE 8.19 Front-line tissue goes to zero cells when removal r exceeds stem-cell proliferation p.

Next, we need to understand how aging can cause the crossing of proliferation and removal rates, namely the failure point. Senescent cells affect proliferation and removal in a way that pushes them toward the threshold (Figure 8.20). Senescent cells secrete SASP that slows down the proliferation of progenitor cells throughout the body. Thus, p is a declining function of X, $p(X)$, Figure 8.20. When senescent cells cross a threshold X_d, proliferation drops below removal, and tissue collapse is predicted to occur. S and D cells vanish. Simulations of the circuit with its feedback loop show how the alveolar cells D go to zero at different times in different individuals (Figure 8.21), according to the time that senescent cells cross the disease threshold (Figure 8.22).

IPF is thus a threshold-crossing disease, and the accumulation of senescent cells with age can induce this threshold crossing. According to our theory, we expect an exponential rise of incidence with age, as senescent cells stochastically cross the disease threshold, with a decline at old age, as is indeed observed (Figure 8.23).

The circuit also explains the clinical observation that the amount of S cells relative to D cells begins to rise before disease onset. This is due to the feedback in the system, which

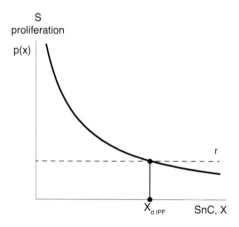

FIGURE 8.20 Senescent cells reduce proliferation pushing it toward removal. At a critical point X_d the tissue collapses.

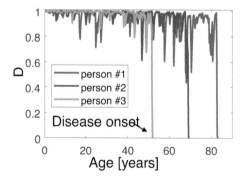

FIGURE 8.21 Cell number drops to zero at different times in different individuals.

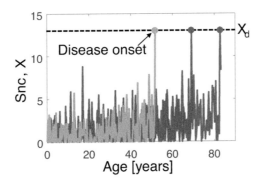

FIGURE 8.22 Disease onset in an individual occurs when senescent cells cross the disease threshold.

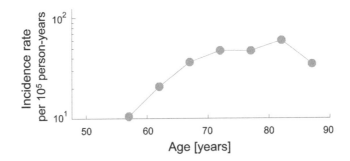

FIGURE 8.23 Incidence of IPF rises exponentially and drops at very old ages.

attempts to ward off the collapse by increasing stem cell numbers. This is a last-ditch attempt to supply the needed number of divisions per unit time to counteract the loss of cells (Solved Exercise 8.2).

Such a threshold for failure is less of a concern in the circuit for protected stem cells, which have low stem-cell removal rate (see Exercise 8.5). Thus, front-line tissues are expected to show age-related fibrotic diseases much more commonly than other tissues.

Now that we understand the origin of the disease threshold, let's also understand the susceptibility to this disease.

SUSCEPTIBILITY TO IPF INVOLVES GENETIC AND ENVIRONMENTAL FACTORS THAT INCREASE STEM-CELL DEATH

Who is susceptible? Most people are not. Their stem-cell proliferation rate is much higher than the removal rate. With age, proliferation rate drops but always stays above removal. The lungs work fine, there is no disease.

But in a small fraction of people, the stem-cell removal rate is higher than in the rest of the population. This is fine at young ages, because proliferation still exceeds removal. But in these individuals, aging can push proliferation down below removal, causing tissue collapse and IPF onset.

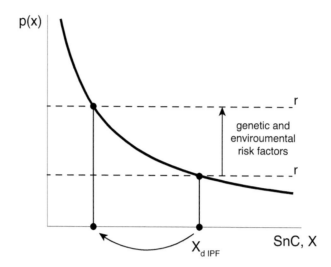

FIGURE 8.24 Individuals susceptible to IPF have a lower threshold due to increased stem-cell removal.

To understand this, we can examine the genetic risk factors for IPF (Martinez et al. 2017). About 15% of IPF cases cluster within families. First-degree relatives of a patient have a 5-fold higher risk of contracting IPF.

There are two classes of gene variants that increase the risk of IPF. The first class is in the surfactant genes expressed by S cells. These variants produce unfolded surfactant proteins that damage the S cells and increase their removal rate r. Increasing cell removal rate lowers the IPF threshold X_d (Figure 8.24). Thus, these gene variants make the disease much more likely.

The other class of genetic risk variants also affects S cells. These are **telomerase** genes. Stem cells have an enzyme called telomerase that allows them to divide indefinitely, by restoring their telomeres after each division. The telomerase risk variants reduce S cell proliferation rate p and increase their death rate r, or equivalently their removal by becoming senescent.

IPF also has environmental risk factors. Smoking doubles the risk of IPF. Smoking is mutagenic, increasing the rate of local senescent cell production, and also increasing removal rates. Exposure to toxins such as asbestos also increases removal and the risk of IPF.

The involvement of elevated removal in IPF also explains why fibrosis begins at the outside of the lung, and then progresses inward. At the outside of the lung, the mechanical stress on the alveoli, and hence removal rate, is highest.

IPF IS MATHEMATICALLY ANALOGOUS TO ANOTHER AGE-RELATED DISEASE, OSTEOARTHRITIS

This theory of IPF can be generalized to other front-line organs, to understand a range of seemingly unrelated diseases. One such disease is a disease of the joints called **osteoarthritis**, a common condition that occurs in about 10% of those over 60 (Martel-Pelletier et al. 2016). In osteoarthritis, the protective cartilage that cushions the ends of the bones wears down over time. The disease most commonly affects joints in knees, hips, hands, and spine. Its symptoms are pain and stiffness in the joints, which can be debilitating. It is a progressive disease with no cure except joint-replacement surgery.

The joint is made of a tough fibrous cartilage. The business end of the cartilage is a smooth surface where the two parts of the joints meet. This is the front line, where the wear-and-tear occurs. The cartilage is constantly remodeled by chondrocyte cells, D, that make the fibers for strength and elasticity, including collagen-2. These D cells are generated by stem-like progenitor cells, S (Koelling et al. 2009). The progenitor cells in the joint are at the front line, just like in the alveoli. The reason is that cells have limited mobility through the cartilage, and thus S cells need to be close to where new D cells are needed, namely at the front line.

The joints suffer mechanical stress, especially in regions that support the body's weight. In the young, this stress doesn't do much and the joints are fine for 50 or more years. But at old age, osteoarthritis can set in. In a process that takes many years due to the very slow turnover of the chondrocytes, D cell number declines, and the fraction of S cells increases. The S cells make tougher fibers than in normal cartilage, such as collagen-1 instead of collagen-2, making the tissue stiffer and less elastic. As a result, cracks form, leading to a hole that often goes right down to the bone.

This hole occurs in the part of the joint that bears the most weight and thus has the highest cell removal rates (Figure 8.25). People with knees that bend inward or outward have the damage at the appropriate side of the knee where load is highest.

Like IPF and virtually all age-related diseases studied so far, removing senescent cells with senolytic drugs alleviates this disease in mice.

Thus, the two diseases IPF and osteoarthritis have a **mathematical analogy.** Stem cells are challenged with a high removal rate because they are at the front line. The removal rate varies across the organ and is highest where the most pressure occurs. Reducing the proliferation rate of S cells down toward their removal rate leads to a rise in the stem-cell

Osteoarthritis is a progressive age-related failure of the joint cartilage.

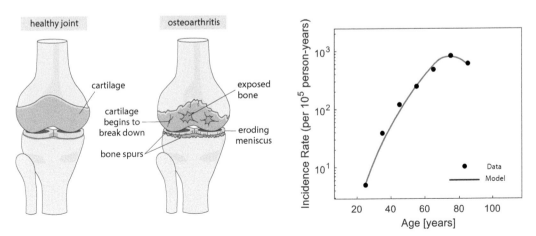

FIGURE 8.25   Osteoarthritis is a progressive age-related failure of the joint cartilage.

fraction S/D and eventually the cells are lost altogether. This reduction in S proliferation can be caused by SASP secreted by the senescent cells in the body, as well as local senescent cells in the joint.

Susceptibility to osteoarthritis, as in IPF, is due to genetic and environmental factors. The main environmental risk-factor for osteoarthritis is being overweight, which increases the load on the joints (Figure 8.26). To see this, note how the higher the body-mass index (BMI, mass divided by height squared), the larger the susceptible fraction ϕ; BMI does not seem to affect the threshold.

Genetic factors are also important, and osteoarthritis has about a 50% heritability. Risk genes include fiber components like certain collagens (including collagen-2) and other cartilage components, as well as gene variants for the signaling molecules IGF1 and TGF-beta relevant to the feedback circuit that helps S and D cells maintain homeostasis.

It is intriguing that diseases as different as a lung disease and a knee disease might have common fundamental origins. In our periodic table in the next chapter, we can expect that other front-line tissues will have similar progressive fibrotic diseases. They form one column in the table.

The disease-threshold model reveals how diseases that seem very different are in fact deeply connected according to the type of threshold that is crossed. Cancer and infectious disease both involve exponential growth when proliferation exceeds removal. Progressive fibrotic diseases occur in the opposite transition, an exponential decline of cells when proliferation of front-line stem cells drops below their removal. When the stem-cell population crashes, the tissue cannot be renewed causing an injury that cannot be repaired.

We are ready to use the disease-threshold model to explore the dynamics of treatment for age-related diseases.

Body mass index (BMI) is a risk factor for osetoarthritis

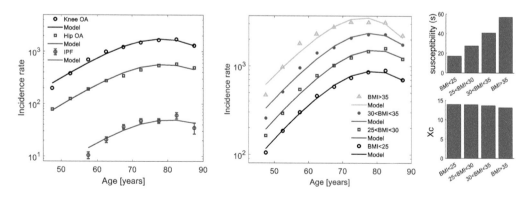

FIGURE 8.26 Risk of osteoarthritis rises with body mass index, which raises susceptibility not disease threshold. Adapted from Katzir et al. (2021).

REMOVING SENESCENT CELLS CAN REJUVENATE THE INCIDENCE OF AGE-RELATED DISEASES BY DECADES

As discussed in the last chapter, age-related diseases are currently treated one at a time. A change of paradigm is to treat them all at once by addressing their core underlying risk factor – aging itself. With our mathematical picture in hand, we can evaluate potential treatments for aging as a core process. We can ask what happens to disease incidence if senescent cells are removed.

In the previous chapter, we mentioned at least three treatment strategies: reduction of senescent cell production by inhibiting the mTor pathway, senolytic drugs that kill senescent cells, and immune therapy that targets senescent cells.

Suppose a 60-year-old starts taking a drug once per month that removes senescent cells. We can simulate this using the saturating removal model by adding a killing term that represents removal of senescent cells due to the drug. Since senescent cells are reduced, they cross the disease threshold at older ages. This predicts dramatic consequences for disease incidence – a rejuvenation on the order of decades. The incidence curve of a typical disease shifts within months to resemble the curve of a younger population (dashed line in Figure 8.27).

Even killing only half of the senescent cells once every month rejuvenates by decades. This works even if we assume, as in Figure 8.27, that senescent cells account for only 25% of the damage responsible for the age-related disease, and the rest is due to currently unknown forms of damage not affected by the drug.

Notably, rejuvenation is predicted even when treatment begins at old ages (Figure 8.28).

Now there was nothing special about the disease we picked for Figures 8.27 and 8.28. Removing senescent cells should similarly reduce the incidence of all age-related diseases.

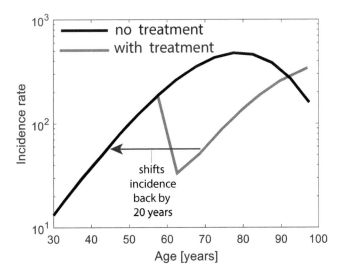

FIGURE 8.27 Removing senescent cells every month rejuvenates the incidence curve by two decades in simulations of the saturating removal model. In these simulations, only 25% of the senescent cells are killed by the drug.

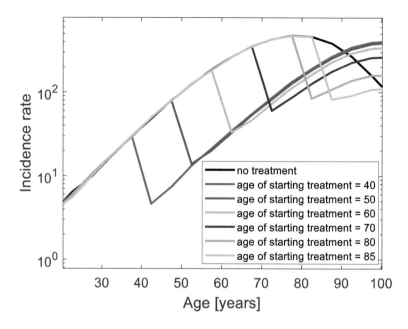

FIGURE 8.28 Treatment starting at old age is predicted to show rejuvenation of incidence curves.

Treating the major risk factor, aging itself, rather than treating one disease at a time can be a turning point in medicine (see Exercise 8.8).

Let's take a nice deep sigh of relief to celebrate. We are ready to sum up the book in a periodic table of diseases.

EXERCISES

Solved Exercise 8.1: Find an analytical form for the incidence curve and age of peak incidence for low susceptibility diseases

The purpose of this exercise is to find an analytical form for disease incidence, using some approximations. It is often useful to derive analytical forms in order to understand the more complex reality.

Let's consider a cohort of susceptible individuals. The disease-free fraction at age t is $F(t)$. The incidence $I(t)$ is given by the number of disease-free individuals of age t that get the disease in the following year. The disease-free individuals are made of those susceptible,

$\phi F(t)$, and those non-susceptible that survive to age t, $(1 - \phi) S(t)$. Thus $I(t) = -\dfrac{\phi \dfrac{dF}{dt}}{\phi F + (1 - \phi) S}$,

where ϕ denotes the susceptible fraction in the population and $S(t)$ is the survival curve.

If we ignore the death rate of the non-susceptible population by setting $S = 1$, and assume that ϕ is small, as it is for most diseases, we obtain $I(t) = -\phi\, dF/dt$.

Let's write incidence in terms of the first-passage-time hazard rate of the disease. The hazard is the number of people per year that get the disease out of the remaining disease-free individuals, $h = -1/F\, dF/dt$. Thus $h = -dlogF/dt$, and $F = e^{-\int_0^t hdt}$

Writing incidence in terms of hazard, we have $I(t) = \phi\, h\, F$, or

$$I(t) = \phi\, h(t)e^{-\int_0^t h dt}$$

Now let's make a simple approximation for the saturating removal model, by approximating the first passage time to cross a threshold goes as the Gompertz law without slowdown, $h = Ae^{\alpha t}$.

We thus find an analytical formula

$$I = \phi A\, e^{\alpha t} e^{-\frac{A}{\alpha}\left(e^{\alpha t}-1\right)}$$

Taking the log of incidence,

$$\log(I) = \log(\phi A) + \alpha t - \frac{A}{\alpha}\left(e^{\alpha t}-1\right)$$

we see a linear rise with a slope α and then a drop at late times when the exponent term becomes large (Figure 8.28). We can also find the time of peak incidence from this equation. Taking $d\log I/dt = 0$ yields $\alpha = A\, e^{\alpha t}$ and thus $t_{\max} = \frac{1}{\alpha}\ln\left(\frac{\alpha}{A}\right)$. The time of peak incidence is inversely related to the incidence slope α, as observed in the data, because the logarithmic term is approximately constant for the range of α in human disease.

Solved Exercise 8.2: Front-line circuits maintain homeostasis using feedback

In order to keep the tissue at homeostasis, and in particular to maintain a proper concentration of D cells, front-line tissues need to have a feedback loop. In this feedback loop, D and S cells signal to each other by secreting molecules that affect differentiation and proliferation rates. If there are too few D cells, for example, these signals act to increase D cell production and restore homeostasis.

In the feedback loop found in the lung and joints, as well as in other stem-cell based organs like the skin, D secretes a signaling molecule that increases S differentiation (one such molecule is $TGF-\beta$). S cells can in principle also secrete factors that increase their differentiation rate. Thus, differentiation rate is a function of D and S concentrations, $q = q(S, D)$.

Let's see how this feedback works. Suppose there is a loss of D cells (Figure 8.29). Since D cells signal to increase differentiation, fewer D cells mean lower differentiation rate q. Thus, at first one makes even fewer D cells. This seems paradoxical. But the reduction in differentiation rate means that more S divisions go to making new S cells instead of D cells. S levels rise, and eventually the larger S cell population supplies more differentiation events per unit time than before the perturbation. D levels rise back. The timescale of this recovery in the alveoli is months, due to the turnover rate of about a month of the D cells (alveolar epithelial cells). In joints, the turnover time is much slower.

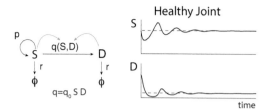

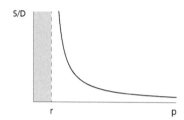

FIGURE 8.29 A front-line tissue with feedback for homeostasis produces damped oscillations of cell numbers.

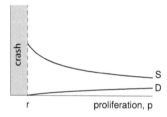

FIGURE 8.30 Fraction of stem cells rises as the disease onset point is approached.

FIGURE 8.31 Differentiated cell numbers reach zero as disease threshold is approached.

This feedback process shows damped oscillations and settles down to a proper steady state. As an aside, we can expect, as in Chapter 3, that such damped oscillations might entrain to the seasons and lead to seasonal changes in alveolar composition, with more S cells and thus more surfactant in some seasons and less in others.

We can also solve the model for various proliferation rates p to observe the rise in S and then the crash as p approaches r. In simulations we use a simple form for the feedback $q(S, D) = q_0 \, S \, D$.

Let's see what happens when the maximal proliferation rate, p, drops to approach the stem-cell removal rate, r. By adding Equations 8.2 and 8.3 the differentiation rate drops out and we remain with $d(S+D)/dt = (p-r)S - rD$. At steady-state, therefore, we have:

$$\frac{S_{st}}{D_{st}} = \frac{r}{p-r}$$

a result that is independent of the nature of the feedback $q(D,S)$. When proliferation rate p drops, the ratio of stem to differentiated cells S/D rises (Figure 8.30) – to keep homeostasis, the feedback loop increases the number of S cells in order to compensate for the reduction in their proliferation rate. The fraction of S cells in the tissue diverges as proliferation p drops toward the removal rate r (Figure 8.30). When $p < r$, both S and D cells reach zero (Figure 8.31).

Exercise 8.3: Stem-cell feedback that keeps constant S.
Consider the following feedback loop. Both stem cells S and D cells secrete factors that increase differentiation rate. The differentiation rate is $q(S, D) = q_0 SD$.

a. Write down the equations for this circuit.

b. Simulate this circuit (or use linear stability analysis) and test whether the steady-state is stable.

c. Show that the steady-state concentration of S cells is independent of S proliferation rate p.

d. What is the concentration of D cells as a function of p?

e. Is the effect of this feedback biologically useful?

Exercise 8.4: Oscillations in front-line tissue circuit.
Consider a feedback loop in which D increases differentiation rate $q(S, D) = q_0 SD$.

a. Write the equations and simulate them.

b. Explain the resulting oscillations in S and D numbers intuitively.

c. Read about the predator-prey model in ecology called the **Lotka–Volterra** model. What is the analogy?

d. Why are ecological population models for species population an interesting resource for modeling cell circuits?

Exercise 8.5: Protected stem cells.
Consider a tissue in which the stem-cell removal rate r_1 is negligible, whereas the D cells have a sizable removal rate r_2.

a. Suppose that a feedback loop provides a stable-steady state. What happens to the S/D ratio as S proliferation p is lowered? Is there a point of collapse?

b. What diseases might characterize such tissues, more often than tissues with stem cells at the front line (high r_1)?

c. Design a feedback loop that provides D levels that are insensitive to variations in stem-cell proliferation p.

Exercise 8.6: NK cell homeostasis circuit.
NK cells are produced by stem cells in the bone marrow. They have a high removal rate r_2, with a lifetime of hours, unless they go into the body's tissues and find cells that make a survival signal (IL15-IL15R). Most cells of the body produce this survival signal. When NK cells touch the donor cells, they receive the signal, and their death rate drops

to zero. NK cells constantly patrol the body and go into and out of the bloodstream and into the tissues.

a. Write equations for NK cell numbers.

b. What determines the NK cell lifetime of about a week in humans?

c. NK cells were introduced into a mouse mutant that cannot produce its own NK cells. These cells lasted for at least 6 months. Explain this result.

d. Explain how this homeostasis mechanism ensures that the number of NK cells matches the number of cells in the tissues that require NK cell surveillance.

Exercise 8.7: Stem-cell symmetric and asymmetric divisions.
Consider the case where a stem cell can divide to form either two stem cells or two differentiated cells, $2S$ or $2D$. This is called symmetric division. Asymmetric division is the case where there is also a third possibility of dividing to produce one D and one S cell.

a. What is the difference in the mathematical equations for the S and D populations in the two cases?

b. How does this affect the S/D ratio as proliferation p approaches removal r_1?

Exercise 8.8: Reducing sickspan and enhancing healthspan
Healthspan is the duration of life spent in good health; the remaining time spent in poor health and disability is called the sickspan. The sickspan is a heavy burden in terms of human well being and economic expenditure. Here we use the saturating removal model to analyze the effect of longevity interventions on the sickspan.

Suppose that poor health begins when senescent cells X cross a threshold X_d, and death occurs at threshold $X_c > X_d$.

a. Use the saturating removal model Equation 7.5 for the mean damage at a given age to calculate the mean healthspan and sickspan.

b. Show that longevity interventions that scale the survival curve increase lifespan and sickspan proportionally. Thus, sickspan is extended. Relative sickspan, namely sickspan divided by lifespan, is unchanged by the intervention. Hint: interventions that scale are interventions that affect damage production rate η.

c. Show that longevity interventions that steepen the survival curve reduce relative sickspan. They extend life mainly by extending healthspan. Hint: these interventions increase β, decrease noise amplitude, or increase both thresholds X_d and X_c.

d. Which type of longevity intervention would you recommend, all things being equal?

e. Senolytics steepen the survival curve whereas caloric restriction scales it. Explain intuitively from biological principles why senolytics might compress the sickspan whereas caloric restriction might extend it by scaling it proportionally to lifespan.

FURTHER READING

Senescent cells and the incidence of age-related diseases. (Katzir et al. 2021)
The Geroscience Hypothesis: Is It Possible to Change the Rate of Aging? (Austad 2016)
A Disease or Not a Disease? Aging As a Pathology. (Gladyshev and Gladyshev 2016)
Pursuing the Longevity Dividend. (Olshansky et al. 2007)

REFERENCES

Armitage, P., and R. Doll. 1954. "The Age Distribution of Cancer and a Multi-Stage Theory of Carcinogenesis." *British Journal of Cancer* 8 (1): 1–12.

Austad, Steven N. 2016. "The Geroscience Hypothesis: Is It Possible to Change the Rate of Aging?" In *Advances in Geroscience*, edited by Felipe Sierra and Ronald Kohanski, 1–36. Springer International Publishing. https://doi.org/10.1007/978-3-319-23246-1_1.

Gladyshev, Timothy V., and Vadim N. Gladyshev. 2016. "A Disease or Not a Disease? Aging As a Pathology." *Trends in Molecular Medicine* 22 (12): 995–6. https://doi.org/10.1016/j.molmed.2016.09.009.

Hernandez-Gonzalez, Fernanda, Rosa Faner, Mauricio Rojas, Alvar Agustí, Manuel Serrano, and Jacobo Sellarés. 2021. "Cellular Senescence in Lung Fibrosis." *International Journal of Molecular Sciences* 22 (13): 7012. https://doi.org/10.3390/ijms22137012.

Katzir, I., M. Adler, O. Karin, N. Mendelsohn-Cohen, A. Mayo, and U. Alon. 2021. "Senescent Cells and the Incidence of Age-Related Diseases." *Aging Cell* 20 (3): e13314. https://doi.org/10.1111/acel.13314.

Koelling, Sebastian, Jenny Kruegel, Malte Irmer, Jan Ragnar Path, Boguslawa Sadowski, Xavier Miro, and Nicolai Miosge. 2009. "Migratory Chondrogenic Progenitor Cells from Repair Tissue during the Later Stages of Human Osteoarthritis." *Cell Stem Cell* 4 (4): 324–35. https://doi.org/10.1016/j.stem.2009.01.015.

Lander, Arthur D., Kimberly K. Gokoffski, Frederic Y. M. Wan, Qing Nie, and Anne L. Calof. 2009. "Cell Lineages and the Logic of Proliferative Control." *PLOS Biology* 7 (1): e1000015. https://doi.org/10.1371/journal.pbio.1000015.

Lopes-Paciencia, Stéphane, Emmanuelle Saint-Germain, Marie-Camille Rowell, Ana Fernández Ruiz, Paloma Kalegari, and Gerardo Ferbeyre. 2019. "The Senescence-Associated Secretory Phenotype and Its Regulation." *Cytokine* 117 (May): 15–22. https://doi.org/10.1016/j.cyto.2019.01.013.

Martel-Pelletier, Johanne, Andrew J. Barr, Flavia M. Cicuttini, Philip G. Conaghan, Cyrus Cooper, Mary B. Goldring, Steven R. Goldring, Graeme Jones, Andrew J. Teichtahl, and Jean-Pierre Pelletier. 2016. "Osteoarthritis." *Nature Reviews. Disease Primers* 2 (October): 16072. https://doi.org/10.1038/nrdp.2016.72.

Martinez, Fernando J., Harold R. Collard, Annie Pardo, Ganesh Raghu, Luca Richeldi, Moises Selman, Jeffrey J. Swigris, Hiroyuki Taniguchi, and Athol U. Wells. 2017. "Idiopathic Pulmonary Fibrosis." *Nature Reviews. Disease Primers* 3 (October): 17074. https://doi.org/10.1038/nrdp.2017.74.

Nordling, C. O. 1953. "A New Theory on the Cancer-Inducing Mechanism." *British Journal of Cancer* 7 (1): 68–72.

Olshansky, S. Jay, Daniel Perry, Richard A. Miller, and Robert N. Butler. 2007. "Pursuing the Longevity Dividend." *Annals of the New York Academy of Sciences* 1114 (1): 11–13. https://doi.org/10.1196/annals.1396.050.

CHAPTER 9

Periodic Table of Diseases

CONGRATULATIONS! WE ARRIVED AT OUR LAST CHAPTER. To celebrate, let's take a nice deep sigh of relief. Here, we tie together the themes and principles of this book in a periodic table of diseases.

PERIODIC TABLE OF DISEASES

As a metaphor, consider Mendeleev's periodic table of elements. In 1869, while preparing a chemistry textbook, Dmitry Mendeleev noticed patterns in the chemical properties of the elements known at the time as a function of their molecular weight (Figure 9.1).

These patterns allowed Mendeleev to predict several new elements (Figure 9.2 white squares). For example, there was an empty space below aluminum that suggested an element similar to aluminum but heavier, with a low melting point and a density of about 6 g/cm^3.

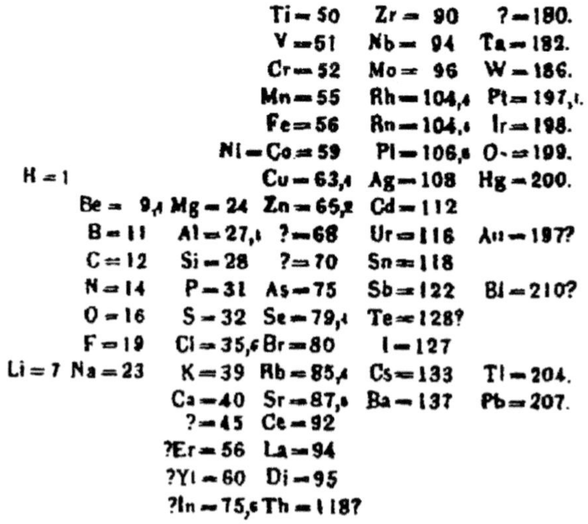

```
                    Ti= 50   Zr = 90   ?=180.
                    V =51    Nb= 94    Ta=182.
                    Cr= 52   Mo= 96    W =186.
                    Mn=55    Rh=104,4  Pt=197,1.
                    Fe=56    Rn=104,4  Ir=198.
                  Ni=Co=59   Pl=106,6  O-=199.
      H =1          Cu=63,4   Ag=108   Hg=200.
          Be = 9,4 Mg = 24  Zn=65,2   Cd=112
          B=11     Al=27,4  ?=68      Ur=116   Au=197?
          C=12     Si=28    ?=70      Sn=118
          N=14     P=31     As=75     Sb=122   Bi=210?
          O=16     S=32     Se=79,4   Te=128?
          F=19     Cl=35,5 Br=80      I=127
      Li=7 Na=23   K=39     Rb=85,4   Cs=133   Tl=204.
                  Ca=40     Sr=87,6  Ba=137    Pb=207.
                  ?=45     Ce=92
                ?Er=56    La=94
                ?Yt=60    Di=95
                ?In=75,6 Th=118?
```

FIGURE 9.1   One of the original versions of Mendeleev's periodic table.

DOI: 10.1201/9781003356929-15

H																	He
Li	Be											B	C	N	O	F	Ne
Na	Mg											Al	Si	P	S	Cl	Ar
K	Ca	Sc	Ti	V	Cr	Mn	Fe	Co	Ni	Cu	Zn	Ga	Ge	As	Se	Br	Kr
Rb	Sr	Y	Zr	Nb	Mo	Tc	Ru	Rh	Pd	Ag	Cd	In	Sn	Sb	Te	I	Xe
Cs	Ba		Hf	Ta	W	Re	Os	Ir	Pt	Au	Hg	Tl	Pb	Bi	Po	At	Rn

FIGURE 9.2 A modern periodic table with Mendeleev's four predicted elements in white squares.

The predicted element, Gallium, was discovered 6 years later with the correct density and melting point. The other three predicted elements were also discovered in due course.

The periodicity of the table remained a mystery for several decades, until quantum mechanics offered the explanation in terms of orbitals.

Metaphors are crucial in science. They let us make inferences about something that is unknown based on something we know. When we say light is a wave or light is a particle, we entail certain features of waves or particles on light. If that seems interesting, I recommend the book *Metaphors We Live By* by Lakoff and Johnson (2008).

So, let's use the periodic table metaphor to organize cell types and diseases. Of course, physiology is much more complex than atoms. The metaphor is imperfect. For example, it won't be a *periodic* table (no repeating period). But there will be patterns and predicted diseases.

CELL TYPES CAN BE CLASSIFIED BY ABUNDANCE AND TURNOVER

Diseases are traditionally classified in four ways: (1) anatomically, by organ or tissue, such as heart diseases and liver diseases; (2) physiologically, by function, such as respiratory diseases and metabolic diseases; (3) pathologically, by the nature of the disease process such as neoplastic diseases (tumors) and inflammatory diseases; and (4) etiologically, by their cause, such as fungal or streptococcal infections.

These classifications are embedded in the international disease code system used to catalog diseases. More recently, diseases have been arranged as networks, where diseases with shared genetic risk factors or symptoms are connected by links (Barabási, Gulbahce and Loscalzo, 2010).

Here, we propose a new classification inspired by the physiological circuit motifs we have studied. The circuits describe the dynamics of cell numbers and cell turnover. We therefore arrange cell types in a table by their number (abundance) and their turnover rate. We will then see how this table reveals patterns of diseases that correspond to each cell type.

Every "element" in the table is thus a cell type (Figure 9.3). The rows go by the number of cells of that type in the body. This ranges from rare cell types, like the parathyroid gland whose size is like a grain of rice with 10^8 cells, to very numerous cell types like the 10^{12} skin cells called keratinocytes that make up the several kilos of skin tissue.

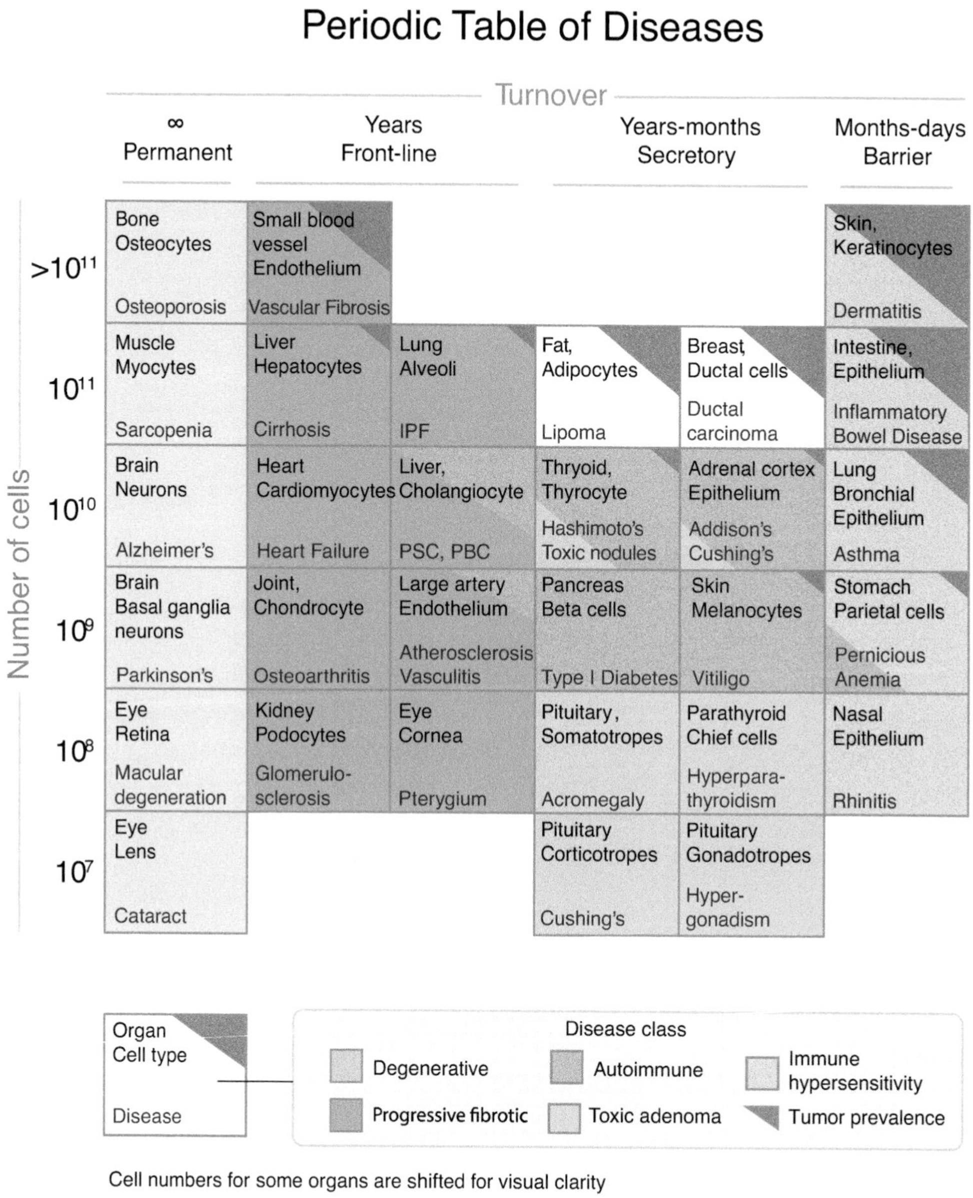

Periodic Table of Diseases

	∞ Permanent	Years Front-line		Years-months Secretory		Months-days Barrier
>10¹¹	Bone Osteocytes Osteoporosis	Small blood vessel Endothelium Vascular Fibrosis				Skin, Keratinocytes Dermatitis
10¹¹	Muscle Myocytes Sarcopenia	Liver Hepatocytes Cirrhosis	Lung Alveoli IPF	Fat, Adipocytes Lipoma	Breast Ductal cells Ductal carcinoma	Intestine, Epithelium Inflammatory Bowel Disease
10¹⁰	Brain Neurons Alzheimer's	Heart Cardiomyocytes Heart Failure	Liver, Cholangiocyte PSC, PBC	Thryoid, Thyrocyte Hashimoto's Toxic nodules	Adrenal cortex Epithelium Addison's Cushing's	Lung Bronchial Epithelium Asthma
10⁹	Brain Basal ganglia neurons Parkinson's	Joint, Chondrocyte Osteoarthritis	Large artery Endothelium Atherosclerosis Vasculitis	Pancreas Beta cells Type I Diabetes	Skin Melanocytes Vitiligo	Stomach Parietal cells Pernicious Anemia
10⁸	Eye Retina Macular degeneration	Kidney Podocytes Glomerulo-sclerosis	Eye Cornea Pterygium	Pituitary, Somatotropes Acromegaly	Parathyroid Chief cells Hyperpara-thyroidism	Nasal Epithelium Rhinitis
10⁷	Eye Lens Cataract			Pituitary Corticotropes Cushing's	Pituitary Gonadotropes Hyper-gonadism	

Legend:
- Organ / Cell type / Disease
- Disease class: Degenerative, Autoimmune, Immune hypersensitivity, Progressive fibrotic, Toxic adenoma, Tumor prevalence

Cell numbers for some organs are shifted for visual clarity

FIGURE 9.3 Periodic table of cell types and diseases. Selected cell types with major non-communicable diseases are arranged according to cell number and cell turnover/tissue function. Disease class is indicated in color. Cell numbers for basal ganglia are shifted for clarity.

To be more tangible, a billion cells weigh about a gram. Exceptions are large cells like neurons, fat cells, and muscle cells which are each 100 times bigger than most cells, so that the 1 kg brain has about 10^{10} neurons.

The columns go by the turnover time of the cell types. Some cell types, like neurons, have no turnover in adulthood. They are called **permanent** cell types. Other cell types, like fat cells, have turnover times of many years. Still others, like liver hepatocytes, turn over with a timescale of a year. Most organs have turnover times of months, with an average cell turnover of about 100 days in the body (Sender and Milo 2021). **Barrier tissues** which stand between the outside and inside typically have the fastest turnover, such as 50d for skin keratinocytes and a few days for the gut epithelium.

The position of a cell type in the table, namely its abundance and turnover, is determined by its physiological function. For example, the size of an endocrine gland is roughly proportional to the size of the target organs. Glands that supply a hormone to the entire body weigh about 10 g, like the thyroid and adrenal. Their 10^{10} cells can make the required amounts of hormones. In contrast, glands that need to supply only a relatively small target organ have fewer cells. Pituitary gland cell types, like those that make ACTH for the adrenal, weigh about 0.01 g, or about 10^7 cells. This is perfect for supplying enough hormone for their 10 g target organ.

The same goes for larger secretory organs. At 1 kg are the liver hepatocytes that produce massive amounts of proteins and metabolites. In the 10 kg range are the skin and fat, all according to their purposes. Skin covers a large area, fat stores fuel.

Turnover is likewise determined by function. Neuron circuits which encode information are permanent and do not replace their neurons. In contrast, barrier cells like skin cells face damage and need to be replaced often.

THE TABLE SHOWS BROAD PATTERNS OF DISEASES

We now consider the main diseases of each cell type, listed in red in each element box (Figure 9.3). The point is that **the diseases show patterns in the table**. Each column has its own class or classes of diseases. These classes are shown in color, corresponding to degenerative diseases, progressive fibrotic diseases, autoimmune diseases, toxic adenomas, and immune hypersensitivity diseases. The orange triangles represent the prevalence of benign tumors and cancers (Figure 9.3).

The first hint of this way of arranging diseases arrived when Yael Korem and I asked which endocrine organs get autoimmune diseases and which do not. We saw a pattern in which organs with less than 10^9 cells get toxic adenomas while organs with more than 10^9 cells get autoimmune diseases, as discussed in Chapter 4. I then began experimenting with the relation between diseases and cell number and turnover around 2019, and the table slowly took. Its current graphical form arose in discussion with several scientists, especially Michael Elowitz.

Is it a table of diseases or cell types? Because we consider cell-type-specific diseases, it is both, where each disease has a cell type, and each cell type can have several diseases.

In this chapter, we will analyze the patterns in the table and their origins in the circuit motifs.

First, however, there are some caveats. The table in Figure 9.3 includes only a small fraction of the cell types in the human body, and only the most prevalent diseases. It is a working draft for us to see patterns. Some classes of diseases do not fit easily into the table. This includes systemic diseases such as lupus and psychiatric diseases such as bipolar disorder, schizophrenia, and depression.

CANCER RISK RISES ALONG THE DIAGONAL OF THE TABLE

To explore the patterns in the table, let's start with cancer prevalence (including benign tumors). The lifetime risk of cancer in a given cell type rises along the diagonal of the table: the more cells and the more exposure, the more mutations in a lifetime and the higher the cancer risk (Figure 9.4).

The main effect is the number of cells. Cells accumulate about 50 mutations per year, regardless of their division rate (Cagan et al. 2022; Miller 2022). Each cell division adds several additional random mutations. The more cells, the more the risk of a cell with oncogenic mutations. The dependence on total cell number is evident in the increasing risk of cancer with height. Each 10 cm of height raises the risk of cancer by a factor of about 1.1. Males have overall 20% more risk of cancer than females, due in part to height difference.

This picture modifies previous notions that the main determinant of cancer risk is the number of lifetime stem-cell divisions (Tomasetti and Vogelstein 2015; Tomasetti, Li and Vogelstein 2017), a theory that could not explain prevalent tumors in slowly dividing tissues. Abundant but slowly dividing cell types get common tumors; these tumors, thankfully, are benign – they don't spread to make metastases and are not lethal. This includes fat cells that have a turnover of years but show lipomas in at least 2% of people. Small blood vessels (capillary endothelial cells) are also very abundant, amounting to 2 kg in a 70 kg person, and get benign angiomas in most people with age.

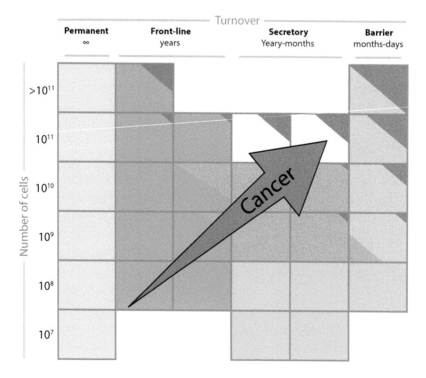

FIGURE 9.4 Risk of cancer, represented by the orange triangles, rises with cell number, cell turnover, and exposure to environmental toxins and mutagens.

Exposure to toxins and mutagens further enhances risk, which is why the risk is highest in barrier tissues – the rightmost column in the table. Barrier tissues are exposed to factors that cause mutations and inflammation, such as UV in the skin, smoking in the bronchi of the lung, and toxins in the gut. They hide their stem cells in a protected niche, as far from the damaging factors as possible. These tissues get prevalent tumors: skin basal cells result in benign tumors in 30% of people; colon and bronchi give rise to colon cancer and lung cancer, each in about 3% of the population; these two currently account for most of the deaths from cancer.

In contrast, permanent tissues cease to divide in adulthood and almost never get cancer in adults; their rare cancers occur in childhood, such as neuroblastomas and osteosarcomas (bone).

SECRETORY CELLS SHOW THREE ZONES: TOXIC ADENOMAS, AUTOIMMUNE DISEASES, AND CANCER

Each column in the table has its own pattern of diseases. Let's start with the column of secretory cells (Figure 9.5), because we have seen it before in Chapter 4.

The secretory cells have a turnover time on the order of weeks to months, except for some cell types like beta cells which have slower turnover in adulthood. Listing the most common diseases of these cell types indicates a striking pattern with three zones of diseases according to cell number.

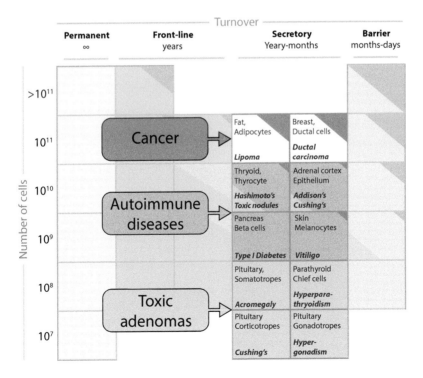

FIGURE 9.5 Secretory cells show three zones: toxic adenomas at the smallest cell numbers, autoimmune diseases between a billion and 10 billion cells, and cancer at higher numbers of cells.

The cell types with the smallest numbers of cells, below 1 g or about 10^9 cells, get toxic adenomas, benign tumors that hyper-secrete the hormone. Examples are parathyroid adenomas that hyper-secrete parathyroid hormone causing excessive calcium, and adenomas of the pituitary that secrete TSH causing hyperthyroidism.

Then, at more abundant cell types, the disease class changes. Cell types between about 1 g and 10 g get cell-type-specific autoimmune diseases in which *T* cells systematically destroy the organ. Examples are type-1 diabetes that destroys beta cells and Hashimoto's thyroiditis that destroys thyroid cells. Instead of too much hormone as in toxic adenomas, these diseases eliminate the endocrine organ and cause too little hormone.

Above 10^{10} cells or 10 g, the most prevalent disease switches to cancer. Cancer cells grow, become de-differentiated and stem-like so they stop their normal function (e.g., stop secreting hormones), and sometimes form metastases that can be lethal.

At the boundaries between the zones we see an overlap of diseases: the 10 g thyroid gets common autoimmune disease (Hashimoto's thyroiditis), but also cancer (thyroid cancer) and toxic adenoma (hot thyroid nodules). The prostate (30 g, not shown in the table) gets cancer and more rarely an autoimmune disease.

THE THREE-ZONE PATTERN CAN BE EXPLAINED BY CIRCUIT MOTIFS

An explanation for this three-zone pattern arises from a shared circuit motif, and from Law 3 – cells mutate, as we saw in Chapter 4. Secretory cells share the "secrete-and-grow" circuit in which a signal controls both secretion and cell growth (Figure 9.6). This circuit is fragile to mutant cells that hyper-sense the signal. Such mutant cells can grow into an adenoma that hyper-secretes the hormone.

The cell types below 1g have so few cells that there is a low probability for hyper-sensing mutants during reproductive years, and even lower probability for the multiple mutations needed for cancer. The strategy is therefore "let it be," with a risk of hyper-secreting adenomas at old age.

The mid-range cell types at 1–10 g have more cells. At birth, there are already numerous cells bearing the hyper-secreting mutations. To avoid hyper-secreting adenomas, we saw that the body may use autoimmune surveillance – the self-attacking T cells we discussed in Chapter 4 – to selectively kill the hyper-secreting cells. The cost is autoimmune disease with a young age of onset in a fraction of the population.

The third zone in this column occurs at more abundant cell types, above 10 g or 10^{10} cells. These cell types do not show autoimmune diseases or toxic adenomas as their main malady. Instead, they show cancer.

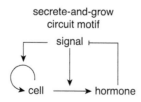

FIGURE 9.6   The secrete-and-grow circuit motif.

One reason for the high cancer prevalence is that at such high cell numbers, one cannot continue to use autoimmune surveillance T-cells because you need so many of these self-attacking T cells that autoimmune disease becomes very likely. There is therefore a switch of strategy. Instead of the differentiated cells making more of themselves as in the thyroid and adrenal (Figure 9.7), these heavier tissues increasingly rely on stem cells (Figure 9.8).

Recall that stem cells are professional dividing cells. They can reduce the number of divisions and hence the number of mutations. The trick is to first differentiate into **transit amplifying cells**: cells which can divide a limited number of times and give rise to the final differentiated non-dividing cell type (Figure 9.8).

For example, if each transit amplifying cell divides 10 times, you get $2^{10} = 1024$ differentiated cells per stem cell division. This amplification reduces the number of divisions and hence mutations in stem cells, the cells that stay in the body for a lifetime. The mutations that arise in the divisions of the transit amplifying cells are not very dangerous, because these cells are soon removed with the natural tissue turnover.

For our purposes, it is important that the stem cell circuit decouples cell division from secretion, and thus does not have the same fragility to hyper-secreting mutant clones as the cells with the secrete-and-grow circuit motif.

Why don't all tissues use stem cells? Stem cells have a cost – risk of cancer. Their stemness provides several hallmarks of cancer even without mutations, such as the ability to divide indefinitely. Stem cells remain in the body and can accumulate the mutations needed for cancer. Thus, one idea is that there exists a tradeoff between hyper-secreting mutants and cancer for cell types in this column; at a certain number of cells, the balance is tipped toward stem cells and cancer.

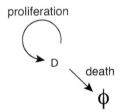

FIGURE 9.7   A cell type which renews itself.

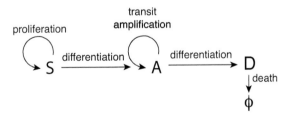

FIGURE 9.8   A cell type which depends on stem cells that differentiate into transit amplifying cells.

THE PERIODIC TABLE EXPLAINS AUTOIMMUNE DISEASES OF NON-ENDOCRINE CELLS

We can begin to use the generalizing power of the table to understand other autoimmune diseases that affect non-endocrine cells, that is cells that do not secrete hormones. We discuss two examples – the diseases multiple sclerosis and ITP. The idea is that cell types in the 10^9–10^{10} "autoimmune zone" with an analogous circuit motif are also fragile to autoimmune disease.

Consider cell types that have a function that needs to be kept under control, and this control is achieved by a circuit analogous to the "secrete-and-grow" circuit. For example, the autoimmune disease multiple sclerosis (MS) attacks brain cells called oligodendrocytes whose role is to coat – to myelinate – neurons in the brain with an insulative wrapping that includes the protein myelin. There are about 10^9 oligodendrocytes in the brain, placing them in the autoimmune zone. Our familiar circuit is at play, in which a "myelinate me" signal from unmyelinated neurons causes the oligodendrocyte precursors to both myelinate and grow (Figure 9.9).

In multiple sclerosis, T cells kill oligodendrocytes, causing severe neurological problems. The main autoantigen is myelin. Multiple sclerosis may thus be a side effect of T-cells that weed out mutant oligodendrocytes that sense too much "myelinate me" signal; such mutant cells would otherwise grow to hyper-myelinate the neurons around them.

A second example is an autoimmune disease that affects blood clotting. In this disease, called immune thrombocytopenic purpura (ITP), T cells kill bone marrow cells called megakaryocytes, also in the 10^9 zone. These are large cells that produce platelets. Platelets are crucial for blood clotting, and purpura eliminates them. This system shows the same circuit motif, with a signal (the hormone thrombopoietin that is degraded by platelets) that makes megakaryocytes both produce platelets and grow (Figure 9.10). Again, the

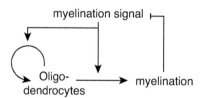

FIGURE 9.9   Oligodendrocytes that myelinate neurons show a circuit in which a signal controls both growth and function, similar to the secrete-and-grow motif.

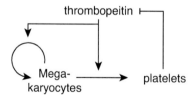

FIGURE 9.10   Megakaryocyte cells produce platelets required for blood clotting. They show a circuit in which a signal controls both growth and function, similar to the secrete-and-grow motif.

autoimmune disease may be a side effect of autoimmune surveillance *T*-cell pruning of hyper-secreting mutant megakaryocytes.

The autoimmune zone of 1–10 g or 10^9–10^{10} cells also extends right and left to some of the other columns, as we will see.

PERMANENT TISSUES HAVE DEGENERATIVE DISEASES OF FAILED MAINTENANCE

Let's go to the column of permanent tissues (Figure 9.11). It includes cell types like neurons, skeletal muscle, bone, and the lens of the eye.

These cells do not divide in adults, but they do have mutations, damage, and continuous maintenance processes. Bone is remodeled at about a teaspoon a day. Neurons are trimmed and repaired by the immune cells of the brain, the glia.

With age, damage production rate rises, due in part to DNA alterations and protein aggregates that accumulate linearly with chronological age. Maintenance capacity is, however, limited. Eventually the trucks of Chapter 7 become overloaded – maintenance processes saturate according to Law 2. In parallel, inflammation rises, further delaying maintenance programs.

These effects lead to diseases of maintenance at old age. Bones show osteoporosis, and neurons show neurodegenerative diseases such as Alzheimer's disease. Skeletal muscles

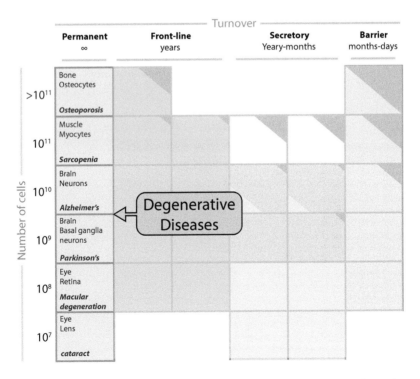

FIGURE 9.11   Permanent tissues show degenerative diseases which are age dependent. Many of these diseases are very prevalent.

show an age-related degenerative disease called sarcopenia in which lean muscle mass is lost at old age at 3%–5% per year if an individual is not active.

Similarly, the lens of the eye shows cataracts. It becomes cloudy, impairing vision, in about 90% of people by age 90. Cataracts are caused by denatured proteins due to slow-down in their removal processes. It is accelerated by diabetes and hypertension, as well as cumulative UV exposure.

The circuit at play summarizes how damage X is produced at a rate that rises with age and saturates its own removal processes, Figure 9.12.

Because of the properties of this circuit that we studied in Chapters 7 and 8, the incidence of degenerative diseases depends strongly on age. Incidence does not, however, depend on organ size: the tiny lens and heavy skeletal muscles both have high prevalence of degeneration. In fact, many of these diseases have very high susceptibility and are part of the shared aging phenotype.

BARRIER TISSUES EXHIBIT IMMUNE HYPERSENSITIVITY DISEASES

At the other extreme, the barrier tissue column, are the cell types with the fastest turnover (Figure 9.13). These tissues stand between the outside world and the inside. This includes the lining of the gut, skin, and lungs.

These organs carry out barrier functions to keep out pathogens and toxins. The barrier functions are modulated by the immune system by means of signal molecules called cytokines. For example, skin thickness is increased upon recurring signals of pathogens or toxins, such as at regions that are repeatedly bruised.

When this regulatory feedback goes wrong, immune hypersensitivity diseases occur. These diseases typically have a young age of onset. An example is psoriasis in which skin cells multiply to cause scaling and inflammation. The inflammation creates positive feedback, with the immune system trying to thicken the barrier even more.

There are three immune hypersensitivity diseases that are often classed together, called the **atopic triad** – asthma in the bronchi, dermatitis in the skin, and atopic rhinitis in the nose (Domínguez et al. 2017). Susceptibility to these diseases often occurs in the same individual, a phenomenon known as **comorbidity**.

This column also includes inflammatory bowel disease (IBD), sometimes called the psoriasis of the gut. The two common inflammatory bowel diseases are Crohn's disease that can occur anywhere in the intestinal tract, and ulcerative colitis (UC), restricted to the colon. These diseases are common, on the order of 1% of the population, with a young mean age of onset.

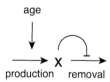

FIGURE 9.12 The circuit for the saturating removal theory of aging. Damage X is produced at a rate that rises with age and saturates its own removal processes.

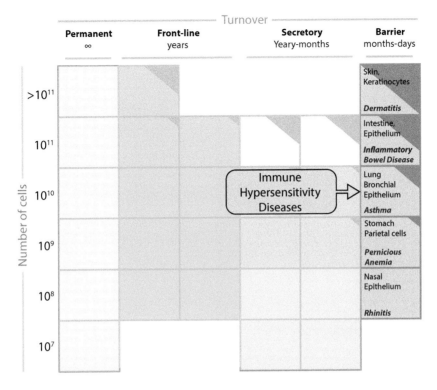

FIGURE 9.13 Barrier tissues show immune hypersensitivity diseases. Most of these diseases have a very young age of onset and can have high prevalence.

The prevalence of these diseases has increased over the last century in high income countries. One hypothesis, called the "old friends" hypothesis, is that improved hygiene has reduced contact with parasites and microbes that used to be common; in the old days, they provided a high "background signal" for the development of the immune system in childhood. Today's low background, due to the lack of these old friends, causes the immune system to develop a more hypersensitive state (Stearns and Medzhitov 2015).

INFECTIOUS DISEASES OCCUR MAINLY IN BARRIER TISSUES

The diseases in the periodic table of Figure 9.3 are non-communicable diseases like cancer and Alzheimer's disease. These account for 9 of the top 10 causes of death in high-income countries, when we include heart disease and stroke caused by the vascular pathology of atherosclerosis discussed below. In contrast, in low-income countries, these diseases account for only 3 of the top 10 causes of death. Infectious diseases are major causes of death, along with neonatal conditions (Figure 9.14).

One can add infectious diseases to the periodic table. There are a vast number of infectious diseases, and again we add only the most prevalent ones that are cell-type specific, or at least have a well-defined cell type as the first site of infection (Figure 9.15). We thus exclude infections that can affect many organs such as staphylococcus and streptococcus infections.

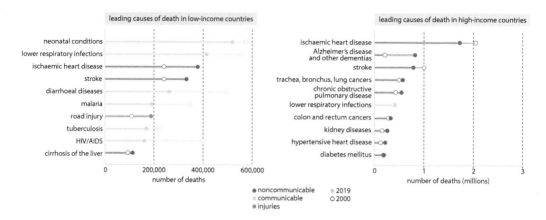

FIGURE 9.14 Leading causes of death in high-and low-income countries. Source: World health organization.

Infectious diseases affect mainly barrier tissues

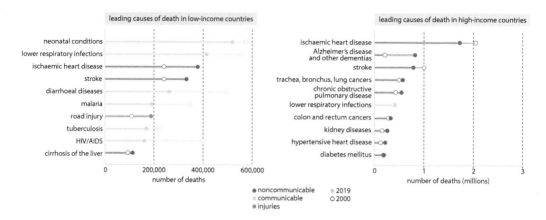

FIGURE 9.15 Infectious diseases affect mainly barrier tissue cell types. Cell types not included in the periodic table of Figure 9.3 are shown as squares below the table.

Most infections occur in barrier tissues. The larger the tissue the more infections, with the skin afflicted with bacteria, virus, fungi, and parasites, and the gut hosting numerous diarrheal diseases and parasites. Other barrier tissues include the urogenital tract with sexually transmitted diseases. We show the urogenital tract and other cell types not in the original table as boxes below the table.

The liver, which is considered part of the digestive system and receives blood directly from the gut, is also a hot spot for parasites and viruses like hepatitis. Many blood-borne diseases like yellow fever and dengue fever target cells of the immune system as their first cell-type hosts. Neurons have specific pathogens like herpes and rabies. Strikingly, most of the internal cell types, including all secretary cell types in the table, lack prominent cell-type-specific infections.

FRONT-LINE TISSUES GET PROGRESSIVE FIBROTIC DISEASES OF OLD AGE

Based on our work in Chapter 8, we devote a column in the table to **front-line tissues** (Figure 9.16). Recall that these tissues have structures that do not allow them to protect their stem cells. The stem cells are at the front line and are removed and damaged about as often as the differentiated cells. We saw the example of the lung alveoli that must be thin for oxygen diffusion and so they can't hide their stem cells.

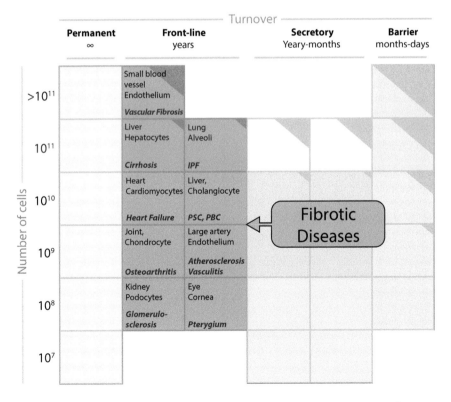

FIGURE 9.16 Front-line cell types, in which stem cells are exposed to damage, show progressive fibrotic diseases. These diseases are age-dependent.

Front-line tissues display age-related progressive fibrotic diseases (Figure 9.16) exemplified by IPF and osteoarthritis discussed in Chapter 8. Progressive diseases develop over time and are generally incurable; end-stage disease is currently treatable only by organ replacement.

These progressive diseases differ from fibrosis after acute injury, which can occur in many organs, especially organs with poor regenerative capacity. An example of acute fibrosis is heart fibrosis after a heart attack. Heart attack is caused by blocked arteries, cutting off the oxygen supply for heart cells and killing them. Another common example of acute fibrosis occurs in the brain after a stroke.

In contrast to acute injury, progressive fibrotic diseases often don't have a clear source of extrinsic damage. As we saw in Chapter 8, the circuit motif at play suggests a common cause for different diseases in this class. They can result from an age-related drop in progenitor proliferation rate below the rate of removal, causing a local collapse of the tissue.

Heart attacks and stroke are often caused by a progressive fibrotic disease of blood vessels, the common age-related pathology of **atherosclerosis**. The arteries are blocked by fatty clots called plaques. The plaques occur in regions of shear stress such as bifurcations of blood vessels, where cell removal rate due to damage is highest. The damage incites inflammation which causes macrophages to accumulate in the lining of blood vessels. Risk of atherosclerosis is increased by factors that damage blood vessels, such as diabetes, smoking, and hypertension. The circuit at play is discussed in Exercise 9.5. Senescence enhances the risk of plaque formation, and the incidence of plaques rises exponentially with age according to the principles discussed in Chapter 8.

There is a hierarchy of blood vessels from the largest arteries and veins with diameters of centimeters to the smallest capillaries with diameters of microns. The smaller the diameter, the more endothelial cells in total in that class of vessels – the small vessels are numerous because they need to fill space. The large arteries show an additional fibrotic pathology – a fibrotic stiffening with age, especially when blood pressure is high. These large arteries are in the 10 g range – the autoimmune disease range – and indeed also get organ-specific autoimmune diseases (vasculitis).

We further include in this column cardiac muscle cells called cardiomyocytes. The heart gets a progressive fibrotic disease leading to **congestive heart failure** when it needs to do excess work over decades, as occurs in chronic high blood pressure. The cardiomyocyte progenitor cells, called satellite cells, are exposed to damage due to the continuous beating of the heart and are thus at the front line.

Other front-line tissues include kidney cells called podocytes. These cells wrap around capillaries and help to filter out waste from the blood and move it into the urine. To function as a filter, they are arranged in a single layer of cells. These cells participate in fibrotic diseases that cause kidney failure. They number about 10^9 cells, placing them in the zone of autoimmunity, and indeed they also exhibit a class of antibody-based autoimmune diseases.

Front-line tissues also occur in the lining of liver bile ducts that is made of secretory cells called cholangiocytes. These cells secrete water and bicarbonate. Cholangiocytes are vulnerable to a progressive fibrotic disease called PSC. They total about 1 g and hence lie also in the zone of autoimmune disease. Cholangiocytes indeed show an autoimmune disease in which *T* cells specifically kill them, called PBC. Thus, this cell type has both an autoimmune disease and a fibrotic disease.

Liver hepatocytes likewise are exposed to toxins since they are the first station for the blood that flows directly from the gut. They are arranged in monolayers sandwiched between blood vessels – a front-line structure. The liver is prone to fibrotic diseases called cirrhosis, as described in Chapter 5. Cirrhosis is caused by damaging agents including viral infection, alcohol, certain drugs, and obesity (fatty liver disease).

Fibrotic diseases often raise the risk of cancer, as in the liver and pancreas, because they supply half of the AND gate: chronic inflammation. Liver cancer almost never occurs without preceding fibrosis.

CIRCUIT MOTIFS UNDERLIE THE DISEASE PATTERNS IN THE TABLE

This is a good point to make a partial summary. Cell-type specific diseases show broad patterns when arranged in a table by cell number and turnover. The disease class in each column is fundamentally due to the kind of circuit motif at play. These circuit motifs act to control organ function and size (Law 1, all cells come from cells) and have fragilities to saturation (Law 2) and mutation (Law 3) that lead to specific types of diseases. It's striking to plot the motifs on the periodic table, as in Figure 9.17.

AGE OF ONSET AND LIFETIME RISK SHOW PATTERNS IN THE TABLE

The table can help to identify additional broad patterns. For example, plotting the mean age of onset of each disease shows a trend in which onset age drops from left to right (Figure 9.18). Onset age thus roughly follows cell turnover time. Degenerative diseases like cataract and Alzheimer's disease peak above age 80, many fibrotic diseases have peak onset at ages 60–70, toxic adenomas at ages 40–60, autoimmune diseases like thyroiditis peak at age 20–40, with type-1 diabetes peaking earlier around age 10. The youngest mean age of onset is for barrier tissues diseases like asthma and dermatitis which often arise in childhood.

Likewise the lifetime risk (susceptibility) shows broad patterns (Figure 9.19). Many degenerative diseases have almost universal susceptibility, 50% to over 90%. Next are barrier diseases with susceptibility on the order of 1%–10%, autoimmune diseases with 0.1%–5%, fibrotic diseases with about 0.1%–1% (except for very common arterial fibrosis and osteoarthritis), and toxic adenomas at 0.01%–1%.

We can also use the table to summarize the major classes of treatment (Figure 9.20). Degenerative diseases like Alzheimer's currently have no effective treatment. Progressive fibrotic diseases at their end stage require organ replacement, such as joint replacement or transplants of heart, kidney, liver or lung. Autoimmune diseases are treated by hormone replacement such as insulin pumps and thyroid hormone supplement. Toxic adenomas are treated by surgical removal. Hypersensitivity is treated by cytokine inhibitors such as anti-TNF antibodies for psoriasis and IBD, as well as glucocorticoids and other anti-inflammatory agents.

Circuit Motifs in the Periodic Table

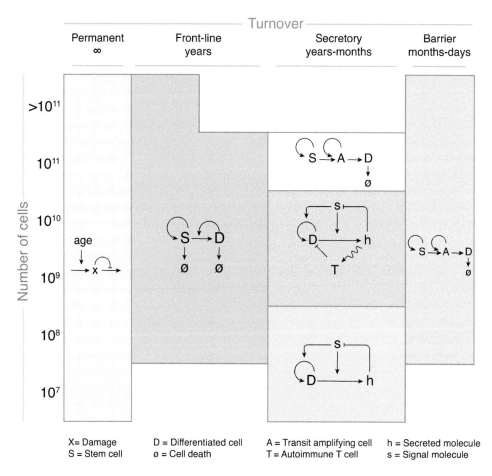

FIGURE 9.17 Each class of diseases in the periodic table corresponds to a specific cell circuit motif.

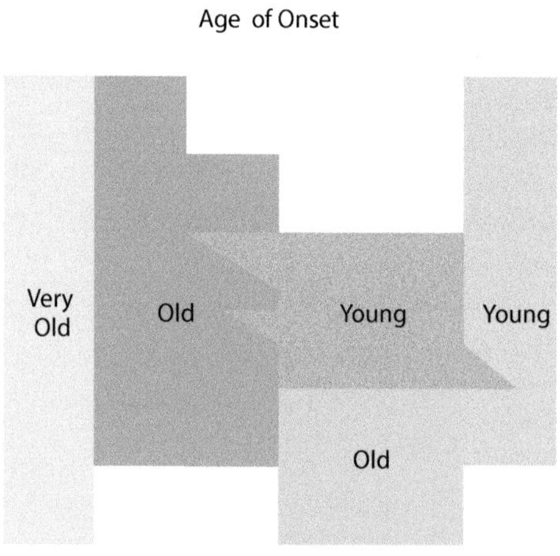

FIGURE 9.18 Age of onset shows patterns in the periodic table of diseases. Disease age of onset from youngest to oldest ranks as immune hypersensitivity, autoimmune, toxic adenoma, progressive fibrotic and degenerative diseases.

Total Lifetime Risk by Disease Class

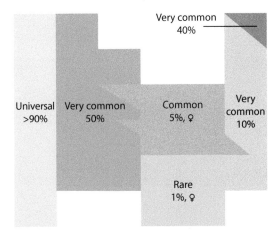

FIGURE 9.19   Disease prevalence shows patterns in the periodic table of diseases. Shown are typical values, some diseases are exceptions.

Current therapy approaches

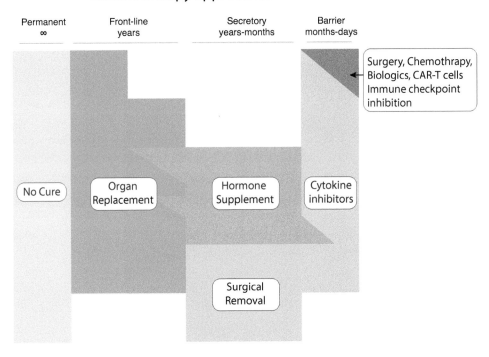

FIGURE 9.20   Current treatments for each class of diseases.

In this book we described several potential treatment strategies for the future (Figure 9.21). Age-related diseases can be treated by senolytics or other approaches that slow the core drivers of aging (Zhang et al. 2022). This is the Geroscience hypothesis, that many age-related diseases can be treated at once by targeting their main risk factor, aging.

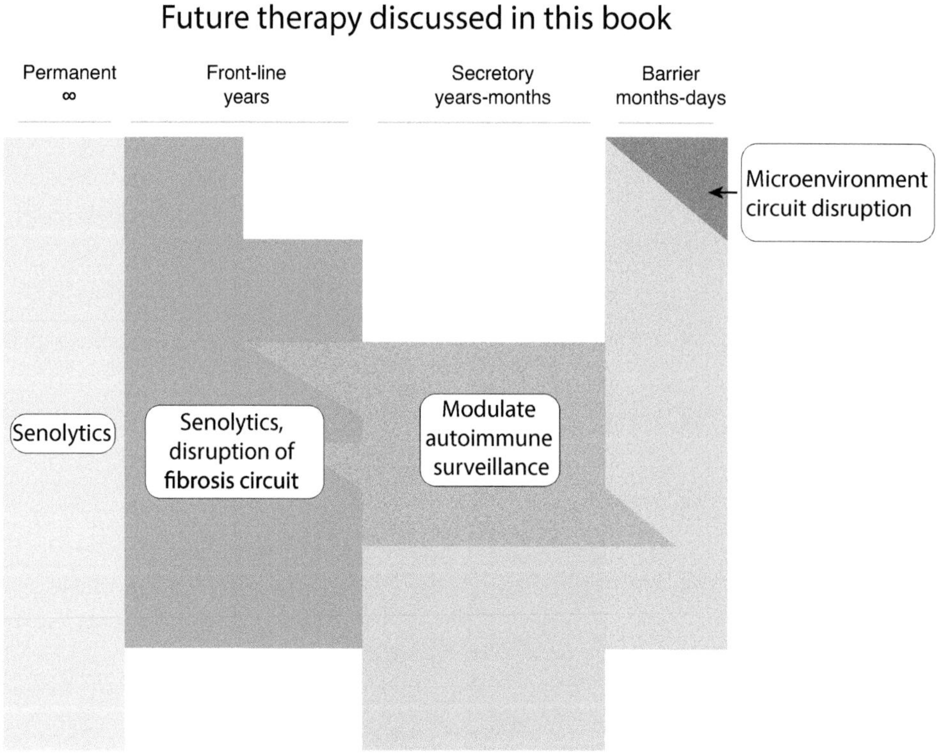

FIGURE 9.21   Possible future treatments for each class of diseases discussed in this book.

Progressive fibrotic diseases may be treatable by disrupting the fibrosis circuit, such as inhibition of the myofibroblast autocrine loop of Chapter 5. This is an example of therapy that aims to modulate ecosystems of interacting cell populations by taking advantage of their most sensitive interactions. An adaptation of this strategy may target the cell populations that make up the cancer microenvironment.

The mutant surveillance theory for autoimmune diseases suggests that the autoimmune surveillance T-cells may have special "license to kill" properties that could make them potential targets for therapy. Therapeutic approaches for a given disease could be adapted for diseases that are adjacent on the table.

PREDICTED DISEASES IN THE TABLE

We can continue with the periodic table metaphor and look for unknown diseases which it predicts.

If we see a secretory cell type in the 1–10 g range, we can predict a *T*-cell-based autoimmune disease. One place to look is in classes of immune cells that secrete alarm signals (Figure 9.22).

For example, plasmacytoid dendritic cells – pDC cells for short – are the main source of the alarm signal interferon-1, which causes cells to raise their defenses against pathogens. The pDC cells sense pathogens using an innate immune receptor (TLR7), which controls their growth and interferon secretion, in a "secrete-and-grow" circuit (Figure 9.22).

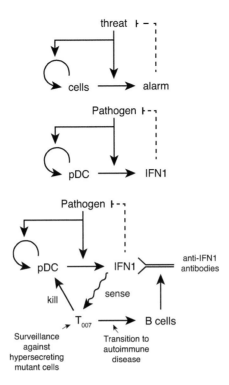

FIGURE 9.22   A predicted autoimmune disease may explain anti-interferon autoantibodies. A circuit in which a signal affects both growth and function of an immune cell is fragile to mutants that hyper-sense the signal. These mutant cells can expand and raise a perpetual alarm. An example is pDC cells that secrete the alarm signal interferon. The autoimmune surveillance theory suggests that *T* cells can remove hyper-secreting mutant pDC cells by sensing an interferon antigen, with the potential to set off a disease in which B cells produce anti-interferon autoantibodies.

The pDC cell type is in the 1g range, predicting an autoimmune disease analogous to type-1 diabetes. The surveillance T-cells should target interferon-1 peptides, and in the predicted disease should activate B cells to make antibodies against interferon-1 (Figure 9.22, bottom panel).

Indeed, about 1% of humanity has anti-interferon-1 antibodies. These individuals have defects in their response to infection, such as an increased risk for severe COVID-19 (Manry et al. 2022).

There may thus be a yet unnamed *T*-cell-based autoimmune disease against pDC, as a fragility of surveillance against hyper-secreting pDC clones that would put the body in perpetual alarm (Figure 9.23).

We may predict analogous autoimmune diseases for other types of immune cells. These diseases may explain the prevalence of anti-cytokine antibodies which are currently a mystery.

Another class of predicted diseases concerns toxic adenomas. If we see a secretory cell type with about 10^8 cells (0.1 g) or fewer, we can guess it might show a hyper-secreting adenoma, a mutant-expansion disease.

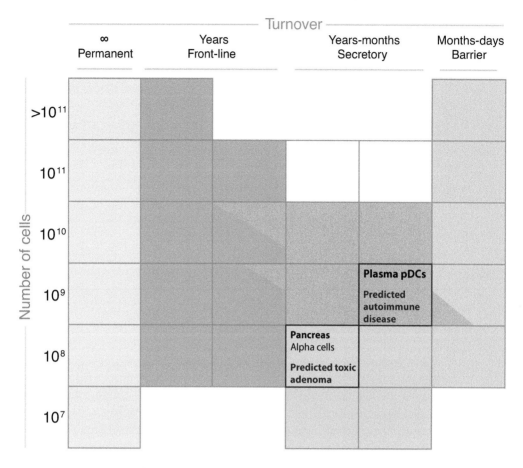

FIGURE 9.23   Examples of predicted diseases in the periodic table of diseases.

As an example, pancreatic alpha cells are in this range. Alpha cells are in the islets together with beta cells. They secrete glucagon, the counter-hormone to insulin, which stimulates the liver to produce glucose out of amino acids, raising blood glucose levels.

The periodic table predicts a mutant-expansion disease of alpha cells at old age, causing excess glucagon, perhaps with a prevalence of around 0.1%. The symptoms should be similar to type-2 diabetes: excess glucose. Such expansions might contribute to diabetes in a small fraction of cases.

A brief technical note: the predicted toxic adenomas are distinct from ultra-rare cancers called neuroendocrine tumors (in this case, glucagonomas). Neuroendocrine cancer is caused by oncogenic mutations that make the cells grow uncontrollably, whereas toxic adenomas remain under control of their size-control circuit. The toxic adenoma cells behave like normal endocrine cells responding to high signal, due to the "delusion" caused by their hyper-sensing mutation.

Another example is a cell-type in the kidneys that secretes renin, a hormone that raises blood pressure. There are about 10^8 such renin-secreting cells in the kidneys, called juxtaglomerular cells. The predicted toxic adenomas should cause hypertension – chronically high blood pressure. They might explain a small part of this prevalent condition.

This discussion raises the possibility of cell-type-specific diseases that are under the radar. These diseases go undetected because their symptoms are subclinical in most conditions. Or perhaps they go undetected because their symptoms are too similar to a common condition like diabetes, hypertension or obesity. The latter possibility suggests ways to find new causes for common pathologies, operative in a small number of cases. Such cases may be unresponsive to the standard treatments for these conditions.

The book is almost at its end! To celebrate, let's take a nice deep sigh of relief. I hope that you enjoyed our journey. We stated basic physiological laws and used them to define circuit motifs. These circuits carry out essential functions like organ size control. They also provide robustness to variations in physiological parameters. However, they have fragilities which are the basis for diseases. The circuits define disease classes that can be organized in a periodic table. I wonder what additional patterns and unifying principles await to be discovered.

EXERCISES

Exercise 9.1: Comorbidity
Suppose that an individual is susceptible to two age-related diseases In the model of Chapter 8, the diseases have thresholds X_{d1} and X_{d2}.

  a. Is there a predicted order in which the diseases occur in a given individual?

  b. Design a computational experiment to discern such ordering in a medical dataset containing the age of onset of diseases in a large number of individuals.

  c. What might be confounding factors in such an attempt to discern ordering?

Exercise 9.2: Disease networks
Diseases can be arranged in networks, where two diseases are linked if they share genetic risk factors or phenotypic features (Barabási, Gulbahce and Loscalzo 2010). Compare this approach to the periodic table. What are its main merits?

Exercise 9.3: Female-male prevalence
Most autoimmune diseases in the secretory cell column are more common in females than males (McCombe and Greer 2014). Most toxic adenomas in the same column are also more prevalent in females. Form a hypothesis for how the autoimmune surveillance mechanism of Chapter 4 may contribute to the origin of a higher female prevalence. Note the fact that many of the endocrine glands involved undergo enhanced cell division during pregnancy and female reproductive cycles.

Exercise 9.4: Biphasic mechanism

We discussed how plasma dendritic cells (pDCs) may harbor a predicted autoimmune disease. These cells show an exhaustion mechanism in which they down-regulate themselves if their receptor is activated for long times (Macal et al. 2018). Discuss this in analogy to biphasic mechanisms like glucotoxicity of Chapter 2.

Exercise 9.5: Rare genetic disorders

Read about cell-type-specific diseases caused by mutations in the fertilized egg, such as cystic fibrosis and Duchenne muscular dystrophy. How would you add them to the table? Do they tend to conform to the disease class of their cell type, or to different classes?

Exercise 9.6: Atherosclerosis master exercise

Read about the mechanisms of atherosclerosis. In atherosclerosis, macrophages and other white blood cells accumulate at a point of inflammation in the artery wall. They recruit smooth muscle cells to proliferate and secrete fibers in a way analogous to myofibroblasts. The macrophages ingest cholesterol-carrying particles (LDL) which get oxidized by the inflammatory environment, and these oxidized LDL particles further stimulate inflammation, macrophage recruitment and macrophage senescence. The upshot is a fibrous and fatty plaque that can grow and eventually block the artery.

a. Write a cell circuit analogous to the fibrosis circuit of Chapter 5.

b. Explain why LDL levels in the blood are a risk factor for atherosclerosis, and derive a mathematical expression for this effect.

c. Explain how senescent cells with their inflammatory factors could push blood vessels across a threshold for atherosclerosis.

d. Explain the exponential incidence of vascular disease with age along the lines of the disease-threshold model of Chapter 8.

REFERENCES

Barabási, A. L., N. Gulbahce, and J. Loscalzo. 2010. "Network Medicine: A Network-Based Approach to Human Disease." *Nature Reviews Genetics* 12 (1): 56–68. doi: 10.1038/nrg2918.

Cagan, A., A. Baez-Ortega, N. Brzozowska, F. Abascal, T.H. Coorens, M.A. Sanders, A.R. Lawson, L.M. Harvey, S. Bhosle, D. Jones, and R.E. Alcantara. 2022. "Somatic Mutation Rates Scale with Lifespan Across Mammals." *Nature* 604 (7906): 517–24. doi: 10.1038/s41586-022-04618-z.

Domínguez-Hüttinger, E., P. Christodoulides, K. Miyauchi, A.D. Irvine, M. Okada-Hatakeyama, M. Kubo, R.J. Tanaka. 2017. "Mathematical Modeling of Atopic Dermatitis Reveals 'Double-Switch' Mechanisms Underlying 4 Common Disease Phenotypes." *Journal of Allergy and Clinical Immunology* 139(6): 1861–1872.e7. doi: 10.1016/j.jaci.2016.10.026.

Lakoff, G., and M. Johnson. 2008. *Metaphors We Live By*. University of Chicago Press. https://books.google.co.il/books?id=r6nOYYtxzUoC.

Lei Dai, Kirill S Korolev, Jeff Gore. 2015. "Relation between stability and resilience determines the performance of early warning signals under different environmental drivers." 112(32):10056-61. doi: 10.1073/pnas.1418415112. PMID: 26216946; PMCID: PMC4538670.

Macal, M., Y. Jo, S. Dallari, A.Y. Chang, J. Dai, S. Swaminathan, E.J. Wehrens, P. Fitzgerald-Bocarsly, and E.I. Zúñiga. 2018. "Self-Renewal and Toll-Like Receptor Signaling Sustain Exhausted Plasmacytoid Dendritic Cells during Chronic Viral Infection." *Immunity.* 48 (4): 730–44. doi: 10.1016/J.IMMUNI.2018.03.020.

Manry, J. et al. 2022. "The Risk of COVID-19 Death Is much greater and Age Dependent with Type I IFN Autoantibodies." *Proceedings of the National Academy of Sciences* 119 (21): e2200413119. doi: 10.1073/PNAS.2200413119/SUPPL_FILE/PNAS.2200413119.SD01.DOCX.

McCombe, P. A. and J. M. Greer. 2014. "Sexual Dimorphism in the Immune System." In *The Autoimmune Diseases*, 319–28. Academic Press. doi: 10.1016/B978-0-12-812102-3.00024-5.

McCombe, Pamela A., and Judith M. Greer. 2020. "Chapter 24 - Sexual Dimorphism in the Immune System." In The Autoimmune Diseases (Sixth Edition), edited by Noel R. Rose and Ian R. Mackay, 419–28. Academic Press. https://doi.org/10.1016/B978-0-12-812102-3.00024-5.

Sender, R. and R. Milo. 2021 "The Distribution of Cellular Turnover in the Human Body." *Nature Medicine* 27 (1): 45–8. doi: 10.1038/s41591-020-01182-9.

Stearns, S.C. and R. Medzhitov. 2015. *Evolutionary Medicine* – Paperback. Oxford University Press. Accessed June 26, 2023. https://global.oup.com/ukhe/product/evolutionary-medicine-9781605 352602

Tomasetti, C. and B. Vogelstein. 2015. "Response." *Science* 347 (6223): 729–31. doi: 10.1126/SCIENCE. AAA6592.

Tomasetti, C., L. Li, and B. Vogelstein. 2017. "Stem Cell Divisions, Somatic Mutations, Cancer Etiology, and Cancer Prevention." *Science* 355 (6331): 1330–34. doi: 10.1126/SCIENCE. AAF9011/SUPPL_FILE/TOMASETTI.SM.PDF.

Zhang, L., L.E. Pitcher, M.J. Yousefzadeh, L.J. Niedernhofer, P.D. Robbins, and Y. Zhu. 2022. "Cellular Senescence: A Key Therapeutic Target in Aging and Diseases." *The Journal of Clinical Investigation* 132 (15): e158450. doi: 10.1172/JCI158450.

Epilogue
Simplicity in Systems Medicine

M Y FAVORITE MOMENTS IN SCIENCE are when you first glimpse simplicity in a complicated system. Simplicity is beautiful. It also provides ways to understand and eventually modulate nature. Finding simplicity is the bread and butter of physics, but no one guarantees simplicity in biology – cells and organs are far more complex than atoms.

That's why this book originates in a sense of wonder. The wonder comes because human biology evolved to function, not to be comprehensible to scientists. There is no *a priori* reason why our immensely complicated physiology and its diseases would be understandable. But still, simplifying principles can be found that make aspects of systems medicine understandable to us.

This epilogue is about simplicity in systems medicine. We will review simplicity in structure and timescales, in the ability to form simplified models, and in the ultimate causes of diseases (Alon, 2003).

SIMPLICITY IN STRUCTURE

One level of simplicity is found in the structure of physiological circuits. There is a huge number of possible ways that cells can regulate each other; the number of possible cell circuits is therefore very large. The surprise is that organs show only a few types of recurring interaction patterns. We call these recurring patterns circuit motifs, because they appear again and again in different organs.

Each circuit motif carries out essential functions, and these functions are universal - they apply to every organ. The first function is organ size control. Without size control, cell numbers would grow or shrink exponentially. The secrete-and-grow circuit motif provides size control by binding together cell growth and cell function in the same signal. This signal, in turn, is regulated by the integrated organ functional output. As a result, the cell population is locked into a size that provides proper function.

Another such circuit motif is the stem-cell feedback motif. Feedback from differentiated cells enhances the rate of differentiation, to keep both stem cells and differentiated cells in homeostatic balance.

These motifs not only determine organ size they also provide a second crucial function: robustness to variations in physiological parameters. Physiological parameters change from person to person and over time. This includes secretion rates, molecular half-lives, blood volume, and sensitivity of tissues to hormones.

The size control motifs compensate for these variations by changing the organ functional mass. This is the principle of dynamic compensation: after a transient period of days to months in which organ size changes, the size reaches a new steady state in which the system behaves as if the parameters had never changed; not only is the steady-state function preserved, but even the temporal response curves to input stimuli are fully compensated.

Basic circuit motifs are sometimes wired together to produce larger circuits (Adler and Medzhitov 2022). For example, two secrete-and-grow motifs are stacked on top of each other in the HPA axis and the thyroid axis. These motifs are wired together in a way that preserves their original functions of dynamic compensation and size control. The larger circuits also have new features, including oscillatory responses on the timescales of months.

SIMPLICITY IN MODELS

In addition to the structural simplicity of a small number of circuit motifs, there is a second level of simplicity. This is the ability to treat circuits with minimal mathematical models that capture the essence of their behavior.

This minimal description is surprising because it contrasts with the complex biochemical mechanisms by which cells carry out their functions. These biological particulars are astoundingly rich, but the dynamics can be described by mathematical models that do not require exhaustive knowledge of the molecular details. The models require only information on whether X activates or inhibits Y, and at what threshold.

In these models, graphical tools like phase portraits and rate plots provide a back of the envelope sketch of the behavior and its dependence on parameters. Even if we do not know the precise form of the interactions, we can deduce the existence of bifurcations in which the system changes behavior all at once. Indeed, we saw that many diseases correspond to such bifurcations – diabetes to the crossing of a glucotoxicity threshold, fibrosis to the crossing of a separatrix in the fibroblast-macrophage phase plane, and age-related diseases to threshold crossing triggered by accumulation of senescent cells.

There is a universal principle that helps such modeling: separation of timescales. Circuit motifs combine slow cell growth on the scale of days to months with much faster molecular signals that work on the scale of minutes to hours. Therefore, one can safely assume that cell numbers remain constant over the time it takes molecular concentrations to reach their steady states.

With this assumption, the models become far simpler. The number of parameters is reduced, because many molecular rates can be lumped together into a single systems-level parameter, often the ratio of molecular production and removal rates. As a result, the essence can usually be captured with models that have only one or two dynamical variables. The qualitative behavior becomes clear in one dimensional rate plots and two-dimensional phase portraits.

Why this simplicity? Evolution may have converged on such understandable circuits because they are the only kinds of designs that can work reliably. If there were three or more dynamic variables essential to the function, and dozens of rate constants with no separation of timescales, the system would be less predictable and can be sensitive to noise inherent in biological systems. An example of such sensitivity is chaos. Chaos can never occur in a two-variable circuit, as guaranteed by a theorem of dynamical systems called the Poincare-Bendixson theorem. You can depend on a simple circuit to predictably do the work, as long as parameters don't cross a bifurcation point.

SIMPLICITY IN ETIOLOGY

In addition to simplicity in structure and in models, there is a third level of simplicity – the ability to discover core drivers of diseases and explain them in terms of basic physiological laws. Some diseases have clear causes or etiology, such as the pathogens in infectious diseases or germline mutations in rare genetic diseases. The origin of many other diseases, however, is complex with multiple interacting genetic and environmental factors. The same goes for aging, which is driven by numerous types of damage. The surprise is that one can sometimes untangle this complexity and predict specific ultimate causes of aging and diseases. These ultimate causes drive the more proximal dysfunctions.

To find simplicity one must look for it, instead of assuming in advance that things are irreducibly complex. We may end up failing, but it is a mistake not to try.

The first conceptual step is therefore to avoid assuming that a disease is an accident of genetics and environment. Instead, it is useful to assume that the disease is due to an unavoidable fragility of a circuit motif. Each disease has a physiological counterpart, a process which protects the circuit from fragility. Disease occurs when the protection mechanism fails because a parameter crosses a threshold, taking the circuit beyond its design specifications. The behavior of the circuit then changes dramatically. Often the change is from stability of cell populations to instability – cell numbers shrink (degeneration, progressive fibrosis) or grow (tumors).

These fundamental causes of many diseases can be deduced from the three physiological laws we discussed. Law 1, all cells come from cells, leads to exponential cell growth that requires size control circuits. Size control circuits, however, are fragile to saturation effects – law 2. When the circuit attempts to compensate for a parameter change, cell mass or other mitigating factor can hit a carrying capacity, abrogating compensation. Homeostasis is lost, as in prediabetes and hypothyroidism.

Size control circuits are also fragile to mutations that inevitably arise according to law 3 – cells mutate. Certain mutations cause a cell to mis-sense its regulatory signal. Such mutant cells mis-interpret a normal level of signal as a high level. These deluded mutant cells behave like normal cells would if the signal was high, and therefore hyper secrete and also outgrow their neighboring cells, threatening to take over the tissue and disrupt homeostasis. As a result, circuits must have protection mechanisms against mutant takeover.

One such mutant-resistance mechanism, biphasic control, programs the cells to kill themselves if their input signal is too high, with the logic that mutant cells that mis-sense the signal are thus eliminated. This, however, adds a fragility to naturally occurring high signal levels, which then can kill all of the cells. This occurs in late stage type-2 diabetes, in which elevated glucose levels lead to glucotoxicity in beta cells.

Another proposed anti-mutant protection mechanism is surveillance by self-reactive T cells. The T cells act as a police force that can weed out hyper-secreting mutant cells. In certain individuals, this surveillance can tip into self-perpetuating autoimmune disease. Examples are type-1 diabetes and Hashimoto's thyroiditis.

The three laws also provide a theory for aging. Law 1, all cells come from cells, applies to stem cells. Law 3 – cell mutate – also applies, leading to stem cells with altered epigenetics that can give rise to damaged progeny cells. The damaged progeny become senescent cells that set off systemic inflammation and reduced regeneration. Law 2 – biological processes saturate – leads to saturation of the removal mechanisms of the damaged and senescent cells. When removal approaches its limit, senescent cell levels rise sharply. They cause systemic inflammation and reduced regeneration that can push the parameters of circuit motifs beyond their operating regime, until a circuit fails – causing the onset of a disease.

For example, reduced regeneration in frontline tissues can cause tissue collapse in susceptible individuals, providing the basis for progressive fibrotic diseases. Rising levels of damaged cells with age also overloads (law 2) the innate maintenance programs that normally remove damaged cells, repair tissues, and fight cancer and infection. This overload gives rise to degenerative diseases in permanent tissues and raises the risk of cancer and infection with age.

These theories of diseases are based on circuit motifs and are thus mathematically tractable. In this way, they are fundamentally different from prose descriptions. The mathematical approach predicts quantitative patterns like the shape of disease incidence curves and the scaling of survival curves. They predict the phases of seasonal oscillations and the months-long trajectories in the aftermath of stress and illness in hormone circuits. Because the theories make quantitative predictions they can be easily falsified. The quantitative predictions can be tested by new experiments and by large biomedical datasets, as we have seen. The more falsification tests they pass, the more confidence we have in the theories.

Thanks to their simplicity, these theories help to pinpoint, out of the sea of potential interactions, those that are likely to be the most promising targets for treatment. For example, the autocrine loop of myofibroblasts stands out as a target for collapsing fibrosis. Experiments inspired by this theory reveal the molecular players in the autocrine loop and inhibit them as described in Chapter 5; as the theory predicts, inhibiting these factors below a threshold reduces fibrosis.

Due to the universality of the theory, the same approach can potentially be translated across organs, with different molecules. For example, inhibiting a myofibroblast autocrine loop inhibits fibrosis in both heart and liver. The same approach might even be translated across pathologies – for example, from progressive fibrosis to cancer microenvironments which also depend on myofibroblast-like cells called cancer-associated fibroblasts.

Finally, simple theories are optimistic. They suggest ways to sculpt the cell populations of the cancer microenvironment, fibrosis, or autoimmunity, and push circuits back across the threshold from disease to health. They point to ultimate causes of aging, which can be targeted to address all age-related diseases at once. Such optimism can help drive new experiments. The results are always more complex and fascinating than the original theory, increasing our understanding. This cycle of theory and experiment is also the seed of novel strategies for treating disease.

REFERENCE

Adler, M. and R. Medzhitov. 2022. "Emergence of Dynamic Properties in Network Hypermotifs." *Proceedings of the National Academy of Sciences* 119(32): e2204967119.
Alon, U. 2003 Biological networks: the tinkerer as an engineer doi: 10.1126/science.1089072

Acknowledgments

A VI MAYO WAS A collaborative, effective, and devoted partner for the figures and computations. Nigel Orme did many of the figures representing anatomy. Ruslan Medzhitov provided inspiration, ideas, and help, and commented on the draft.

I'm grateful to my former graduate students whose research is part of the backbone of this book. Omer Karin, when did his PhD with me, was a pioneer in our transition from systems biology to systems medicine. Many of the principles in this book arose in our joint work, mainly biphasic mutant resistance and dynamic compensation of Chapters 2 and 3 and the saturating removal model of aging in Chapters 6 and 7. Miri Adler and Shoval Miyara worked on fibrosis as bistability in Chapter 5, Alon Bar and Avichai Tendler on seasonality in Chapter 3, Tomer Milo and Lior Maimon on bipolar disorder in Chapter 3, Yael Korem Kohanim on the thyroid axis in Chapter 3 and the surveillance theory of autoimmune in Chapter 4, Itay Katzir on the incidence of age-related diseases in Chapter 8, Pablo Szkeley on the mass longevity triangle in Chapter 6, and Yifan Yang on the saturating removal model in Chapter 7. Avi Mayo participated in all of these. Additional research input came from Aurore Woller, Yuval Tamir, Moria Raz, Michal Shilo, Hila Sheftel, Tomer Landsberger, and David Glass.

Work with colleagues informed the book as well, with Ruslan Medzhitov in Chapters 4 and 5, Johannes Dietrich Chapters 2, 3, and 4, Dan Jarosz, Nan Hao, Lev Tsimring Jeff hasty Chapters 6 and 7. Amos Tanay and Neta mendelssohn Cohen provided Clalit data for Chapters 3, 8, and 9. Valery Krizhanovsky and Amit Agrawal for senescence cell work in Chapter 7. Stefan Kallenberger, Scott Friedman, Shuang Wang, Eldad Tzahor, Shimrit Mayer, and Ruth Scherz-Shouval for fibrosis experiments in Chapter 5.

I wrote much of this book during a sabbatical at Stanford and the CZI biohub. Steve Quake was the perfect host. Steve and Liquon Liu took the course based on the book draft at Stanford bioengineering and provided great insights. Other colleagues who read the draft or otherwise helped include Sisi Chen, Bo Wang, James Ferrell, Eytan Yaffe, Mike Grecius, Ed Marti, Greg Huber, Thea Tlsty, David Schnieder, Reviel Netz, and Wendell Lim.

My brother Gidi Alon was a superb partner to craft the narrative of early Systems Medicine lectures, and my niece Roni Stok was an active participant in two courses, Systems Medicine and Hormone circuits. Gefen Alon and Tamar Alon helped with song lyrics. Galia Moran was always encouraging and supportive.

I thank the Weizmann Institute students in Systems Medicine 2019, 2020 and in Hormone Circuits 2021, and Stanford students in BE333 2021, whose questions helped me to be more clear. Thank you to CRC press and Elliot Morrisa who kept me oriented.

Michael Elowitz helped to design the graphics for the periodic table in Chapter 9. He hosted me, with Barbara Wold, to give a four lecture nanocourse at Caltech to test the book out, which gave me a boost of motivation. Michael also read the entire book draft and edited it. Some of his comments were hilarious.

Index

Note: **Bold** page numbers refer to tables and *italic* page numbers refer to figures.